ADVANCES IN SPACE RESEARCH

The Official Journal of the Committee on Space Research (COSPAR)
A Scientific Committee of the International Council of Scientific Unions (ICSU)

VOLUME 19, NUMBER 1

RADIATION FROM BLACK HOLES, FUTURE MISSIONS TO PRIMITIVE BODIES AND MIDDLE ATMOSPHERIC FINE STRUCTURES

Symposium E1.1

Sponsors

COMMITTEE ON SPACE RESEARCH (COSPAR)
INTERNATIONAL ASTRONOMICAL UNION (IAU)

Program Committee

N. White, U.S.A.
G. Hasinger, Germany
C. Day, U.S.A.
M. Elvis, U.S.A.
R. Sunyaev, Russia
Y. Tanaka, Japan
A. C. Fabian, U.K.

Symposium B1.5

Sponsors

COMMITTEE ON SPACE RESEARCH (COSPAR)

Programme Committee

G. H. Schwehm, Netherlands
T. Mukai, Japan
J. Rahe, U.S.A.

Symposium C2.4

Sponsors

COMMITEE ON SPACE RESEARCH (COSPAR)

Programme Committee

E. Kopp, Switzerland
C. R. Philbrick, U.S.A.

RADIATION FROM BLACK HOLES, FUTURE MISSIONS TO PRIMITIVE BODIES AND MIDDLE ATMOSPHERIC FINE STRUCTURES

Proceedings of the E1.1, B1.5, and C2.4 Symposia of COSPAR Scientific Commissions E, B and C which were held during the Thirtieth COSPAR Scientific Assembly, Hamburg, Germany, 11–21 July 1994

Edited by

C. DAY (E1.1)

Laboratory for High Energy Astrophysics, NASA Goddard Space Flight Center, Code 660-2, Greenbelt MD 20771, U.S.A.

T. MUKAI (B1.5)

Department of Earth and Planetary Science, Faculty of Science, Kobe University, Nada, 657 Kobe, Japan

G. SCHWEHM (B1.5)

European Space Agency, ESTEC, Postbus 299, NL2200 AG Noordwijk, The Netherlands

and

C. R. PHILBRICK (C2.4)

Department of Electrical Engineering, Pennsylvana State University, 121 Electrical Engineering East, University Park, PA 16802-2705, U.S.A.

Published for

THE COMMITTEE ON SPACE RESEARCH

PERGAMON

U.K.

Elsevier Science Ltd, The Boulevard, Langford Lane,
Kidlington, Oxford OX5 1GB, U.K.

U.S.A.

Elsevier Science Inc., 660 White Plains Road,
Tarrytown, New York 10591-5153, U.S.A.

JAPAN

Elsevier Science Japan, Tsunashima Building Annex,
3-20-12 Yushima, Bunkyo-ku, Tokyo 113, Japan

First edition 1997

ISBN 0-08-043097-X

In order to make this volume available as economically and as rapidly as possible the author's typescript has been reproduced in its original form. This method unfortunately has its typographical limitations but it is hoped that they in no way distract the reader.

Whilst every effort is made by the publishers and editorial board to see that no inaccurate or misleading data, opinion or statement appears in this journal, they wish to make it clear that the data and opinions appearing in the articles and advertisements herein are the sole responsibility of the contributor or advertiser concerned. Accordingly, the publishers, the editorial board and editors and their respective employers, officers and agents accept no responsibility or liability whatsoever for the consequences of any such inaccurate or misleading data, opinion or statement.

NOTICE TO READERS

If your library is not already a subscriber to this series, may we recommend that you place a subscription order to receive immediately upon publication all new issues. Should you find that these issues no longer serve your needs your order can be cancelled at any time without notice. All these conference proceedings issues are also available separately to non-subscribers. Write to your nearest Elsevier Science office for further details.

Contents Direct delivers the table of contents of this journal, by e-mail, approximately two to four weeks prior to each issue's publication. To subscribe to this free service, complete and return the form at the back of this issue or send an e-mail message to cdsubs@elsevier.co.uk

Transferred to digital print 2009
Printed and bound in Great Britain by CPI Antony Rowe, Chippenham and Eastbourne

HIGH ENERGY RADIATION FROM GALACTIC AND EXTRAGALACTIC BLACK HOLES

Proceedings of the E1.1 Meeting of COSPAR Scientific Commission E which was held during the Thirtieth COSPAR Scientific Assembly, Hamburg, Germany, 11–21 July 1994

Edited by

C. DAY

Laboratory for High Energy Astrophysics, NASA Goddard Space Flight Center, Code 660-2, Greenbelt MD 20771, U.S.A.

CONTENTS

FUTURE SPACE MISSIONS TO PRIMITIVE BODIES

FINE STRUCTURES IN THE MIDDLE ATMOSPHERE AND THEIR ORIGIN

APPENDIX

 Pergamon

PII: S0273-1177(97)00109-9

Adv. Space Res. Vol. 19, No. 1, p. (1)1, 1997
© 1997 COSPAR
Printed in Great Britain. All rights reserved
0273–1177/97 $17.00 + 0.00

EDITORIAL COMMENT

This is the last issue of *Advances in Space Research* devoted entirely to the papers presented at the Hamburg COSPAR assembly held in 1994.

Most of this volume is devoted to papers in session E1.1 titled High Energy Radiation from Galactic and Extragalactic Black Holes edited by Dr C. Day. Unfortunately the original manuscripts submitted by Dr Day in 1995 vanished in the postal service. Equally unfortunately, the loss of these manuscripts was not discovered for several months. Dr Day has contacted the authors of papers in this session and they have resubmitted their papers for publication. This is the final section of Commission E papers from the Hamburg assembly to be published.

The papers from Session B1.5, Future Space Missions to Primitive Bodies, and Session C2.4, Fine Structures in the Middle Atmosphere and Their Origin, do not have enough papers to constitute a single volume of *Advances in Space Research*. These papers have been included in this issue.

In keeping with the decision of the COSPAR Publication Committee to publish late papers in *Advances in Space Research*, the Appendix of this issue contains several papers that were either received too late for inclusion in an earlier publication or were misdirected in the editorial process. The paper by Igarashi *et al.*, was presented in the C.3 Session on the Southern Hemisphere Upper Atmosphere and Ionosphere. Readers wishing to read the papers already published from this session are referred to *Advances in Space Research*, Vol. 16, No. 5 .

The final two papers in the Appendix are from the Washington, DC COSPAR Assembly and World Space Congress held in 1992. These papers were presented in the B6-M session on the Shapes and Gravitational Fields of the Planets and Smaller Bodies of the Solar System.

M. A. Shea
Editor-in-Chief

Pergamon

Adv. Space Res. Vol. 19, No. 1, pp. (1)5–(1)14, 1997
© 1997. Published by Elsevier Science Ltd on behalf of COSPAR
Printed in Great Britain. All rights reserved
0273–1177/97 $17.00 + 0.00

PII: S0273-1177(97)00030-6

X-RAY SPECTRAL PROPERTIES OF CYG X-1

Ken Ebisawa

Code 668, NASA/GSFC, Greenbelt MD 20771, U.S.A., and Universities Space Research Association

ABSTRACT

Cyg X-1 is a prototype of black hole candidates with fiducial evidence of large mass. In this paper, I review X-ray spectral properties of Cyg X-1. Usually, Cyg X-1 is in the low state (hard spectral state), in which the energy spectrum in 2 – 10 keV is roughly represented with a power-law function, and sub-second short-term variations are significant. The continuum of the low state energy spectrum is consistent with the superposition of a power-law function with an exponential cut-off and its reflection component by an optically thick accretion disk. Thermal Comptonization of soft photons by hot plasma is a likely model for the intrinsic power-law emission. Iron K-emission lines have been reported, and study of those emission lines have a potential to verify strong gravitational fields around a black hole. However, the line emission is not prominent, as opposed to the cases of AGNs, and the line parameters are dependent on the choice of the continuum model. To determine the continuum spectral shape precisely and disentangle the iron line emission, observations having both high spectral resolution and wide energy coverage will be required. *ASCA* and *XTE* simultaneous observations may provide such a chance.

INTRODUCTION

Since early days of the X-ray astronomy, the extraordinary characteristics of Cyg X-1 have been attracting researchers' interest, and quite a few observations have been carried out by numerous high energy astronomical missions. Cyg X-1 is probably a black hole binary, having evidence of the massive central object ($\gtrsim 9.5$ $M_\odot$/1/; see also /2, 3/ for early reviews on Cyg X-1). Today, about a dozen of X-ray binaries are designated as black hole candidates having the evidence of large mass and/or X-ray properties shared with Cyg X-1. For recent reviews on galactic black hole candidates in general, see, e.g., /4, 5, 6, 7, 8/. Cyg X-1 is a prototype of black hole candidates, and studying Cyg X-1 is expected to lead to the general understanding of black hole candidates. In this paper, I shall review X-ray spectral properties of Cyg X-1 mainly based on published material. In addition, *Ginga* re-analysis results and preliminary *ASCA* results are presented.

BIMODAL STATES

Cyg X-1 has distinct bimodal states, namely, the low state (hard spectral state) and the high state (soft spectral state), so do other black hole candidates such as GX339–4 (e.g.,/9/) and GS1124–68/GRS1124–68 /10, 11/. Usually ($\gtrsim 90$ % of the time), Cyg X-1 is in the low state in which the 2 – 20 keV luminosity is lower than that in the high state and the energy spectrum is approximated by a power-law function with the photon-index ~ 1.7. So far, the high state has been observed only from 1970 to 1971, in 1975 and in 1980 (see references in /2, 3/). Pivotal spectral changes were reported between the two spectral states such that the soft X-ray luminosity below ~ 8 keV is higher in the high state, while the hard X-ray luminosity is lower (e.g., /12/).

Although observations of the Cyg X-1 high state are so poor, some other black hole candidates commonly show the high state, and high state spectral characteristics have been extensively studied. The high state energy spectra comprise two components, namely the (ultra-)soft component and the hard component. The soft component, whose characteristic temperature ($\lesssim 1$ keV) is significantly lower than that of neutron star binaries, is considered to be from optically thick accretion disks, and its luminosity and spectral changes are described simply by the change of the mass-accretion rate, with the inner disk radius being invariant /10, 11, 13, 14, 15, 16/. A precise disk spectral calculation taking account of the radiative transfer has shown that the mass of the central object in LMC X-3 estimated from a model fitting to X-ray energy spectra is consistent with the dynamical mass derived from optical observations /17, 18/.

The hard component in the high state is roughly expressed with a power-law function with a high energy cut-off, and often extends as far as several hundreds keV (e.g., /19, 20/). The hard component is more variable than the soft component in various time-scales, and the variation appears to be independent of the soft component /11, 13, 15, 21/. Probably, the optically thin region close to the black hole, inside or surrounding the optically thick accretion disk, is responsible for the production of the hard component, and similar mechanisms for the low state energy spectra (see the next section) will be working.

The origin of the bimodal transition is not completely elucidated; change of the mass accretion rate may trigger the transition (e.g., /22, 23, 24/), and state transitions of Cyg X-1, GX339-4 and GS1124-68 may be explained in this model. However, there are at least three X-ray novae which obviously do not fit in this scheme. GS2000+25 exhibited the high state features from the beginning to the quiescence /13/. On the other hand, GS2023+33/13/ and GRS 1716-249 (= GRO J1719-24)/25/, when discovered at very luminous states, exhibited the low state characteristics, namely, power-law like energy spectra and rapid short-time variations.

We will be able to study the next Cyg X-1 high state in a greater detail than before, having several superior high energy instruments together (let's hope the next high state occurs during the time *ASCA*, *GRO* and *XTE* are in orbit simultaneously !). The low state energy spectra of Cyg X-1 have been extensively studied by many high energy missions. In the following sections, all the spectral characteristics are concerning the low state.

CONTINUUM EMISSION

The continuum emission is roughly approximated with a power-law function with an exponential cut-off at ~ 100 keV, and consistent with the Comptonized spectra of soft photons by hot plasma with $T_e \sim 27 - 60$ keV and $\tau \sim 1 - 5$ (/26, 27, 28, 29/, see also references in /3/). Note that these plasma parameters were obtained by applying optically thick Comptonization model by Sunyaev and Titarchuk (1980) /27/, and may not be correct, since there is evidence that the Comptonizing cloud is optically thin (see below). Copious soft photons are presumably from the optically thick accretion disk, and maybe additionally from cyclotron emission inside the accretion disk /30/. The inner-region of the accretion disk may be optically thin and geometrically thick, such that the electron temperature is hot enough to Comptonize the soft photons up to hard X-ray energies/22, 31/. Alternatively, the Comptonizing hot corona may be surrounding the optically thick accretion disk, and the corona is heated by, e.g., dissipation of magnetic energies /32, 33, 34/.

If the source is sufficiently compact and luminous in high energies, as is the case for black hole candidates and AGN, $e^{\pm}$ pair production will become significant and play a substantial role in producing emergent spectra. In the fall of 1979, a strong gamma-ray bump was observed in 0.5 – 2 MeV /37/, and this was interpreted as due to emission from a very hot (~ 400 keV) pair-dominated plasma /38, 39/. Such transient emission features, which may be associated with the 511 keV annihilation line, have been observed also from the black hole candidates 1E1740.7–2942 and GS1124–68/GRS1124–68, and the Crab nebula/40/[1].

A disk reflection feature has been observed (see the section below) as a superposition on the power-law spectrum, and this suggests the Comptonizing cloud is optically thin, since optically thick clouds would smear out any local spectral feature such as absorption edges in the reflection component. Haardt et al. (1993) /41/ claimed that the electron temperature and optical depth are constrained from 2 – 20 keV observations, since the reflection spectrum is dependent on the optical depth and temperature of the scattering medium. Applying their disk corona model taking

---

[1]These features may be the 478 keV line associated with the reaction $\alpha + \alpha \to Li^7 + p + \gamma$. See Sunyaev et al. in this issue.

account of the disk reflection /42, 43/ to *EXOSAT* GSPC data, they have derived the plasma parameters $T = 140 - 190$ keV and $\tau = 0.26 - 0.45$. Although these parameters are very different from those obtained by applying Sunyaev and Titarchuk (1980) /27/ optically thick model to *OSSE* and *SIGMA* high energy spectra /28, 29/, Haardt et al. (1993) argued that their best-fit model to Cyg X-1 has a similar spectral shape to the Sunyaev and Titarchuk (1980) model above 20 keV, and can explain these high energy data as well. They claimed that the plasma parameters obtained with the Sunyaev and Titarchuk (1980) model are not correct, since the optically thick Comptonization model was applied to the composite spectra of the emission from optically thin plasma and its reflection component.

Recently, Titarchuk and his colleagues published a new Comptonization model /44, 45, 46/ as a sophistication of the Sunyaev and Titarchuk (1980) model. In their new model, relativistic effects are taken into account (Klein-Nishina cross section is used instead of the Thomson cross section) and the formula is generalized into the optical thin regime. Consequently, the new model is valid for optically thin emission from high temperature plasmas ($\gtrsim 100$ keV), as well as optically thick emission with lower temperatures. Titarchuk (1994) /44/ showed his model with similar plasma parameters ($T = 150$ keV and $\tau = 0.15$ or 0.62 according to the geometry) and Haardt et al. (1993)'s best-fit model to Cyg X-1 yield very similar energy spectra above 20 keV. Note that Haardt et al. (1993) model does not take into account the relativistic correction, and Titarchuk (1994) model does not include the disk reflection. Probably both effects will be important to describe realistic energy spectra.

IRON K-LINE/EDGE SRUCTURE AND DISK REFLECTION

Gas scintillation proportional counters (GSPC) on-board *Tenma* and *EXOSAT* had significantly better energy resolution (FWHM ≈ 8 % at 6 keV) than standard proportional counters, and could carry out X-ray spectroscopy of Cyg X-1 practically for the first time. Barr et al. (1985) reported the discovery of a *broad* (FWHM ~ 1.2 keV) emission line centering at ~ 6.2 keV with an equivalent width of typically 120 eV from the *EXSOT* GSPC spectra /47/. They argued the line is Compton red-shifted iron K-line from optically thick ($\tau > 5$) accretion disk corona. Fabian et al. (1989) showed the possibility of the fluorescent line emission from inner region of the optically thick accretion disk /48/. In this model, broad line-width is explained mainly by the effects of Doppler-broadening, and transverse and gravitational redshifts are also significant. They pointed out the expectation of measuring the black-hole mass and disk inclination, in near future, from precise iron line observations (see also /49/). Kitamoto et al. (1990) /50/, on the other hand, could fit their *Tenma* GSPC spectra with a *narrow* emission line at around 6.5 keV with an equivalent width of 60 – 80 eV. The low-energy absorption column density was clearly variable with orbital phases, and there was also a hint of line orbital variations. They argued the line may be due to fluorescence in the stream of matter from the companion and/or in the outer, extended parts of the accretion disk, being subject to the orbital variations.

Note that the continuum was modeled by a single power-law function and iron K-edge was not taken into account both in Barr et al. (1985) and Kitamoto et al. (1990), but this is probably incorrect (see below). Since the line feature is very dependent on the modeling of the continuum shape and edge structure, the line parameters obtained by both Barr et al. (1985) and Kitamoto et al. (1990) should be appraised with extreme care.

The *Ginga* satellite /51/ was launched in 1987, and the on-board LAC detector /52/ had a 4,000 cm^2 effective area in 2 – 37 keV (or 2 – 60 keV by changing the gain), which is the largest ever until *XTE* is flown. *Ginga* found energy spectra of Cyg X-1, as well as of other black hole candidates, have such clear deviations from a power-law function that look like a broad absorption feature in 7 – 20 keV and a hump above 20 keV /8, 14, 53, 54, 55/. The feature, though obvious with *Ginga*, had not been explicitly noticed before, since the absorption feature is not sharp, and previous missions did not have enough statistics above ~ 15 keV^2.

---

[2]Bałucińska and Hasinger (1991) /56/ noticed the deviation from the power-law model at $\gtrsim 10$ keV in their *EXOSAT* ME data (see their figure 5a), but they ascribed the deviation to the systematic calibration uncertainty at

The broad absorption in 7 – 20 keV and a hump in $\gtrsim$ 20 keV is considered due to reprocessing and reflection of X-rays by an optically thick accretion disk (/57, 58/). The reflection component has an iron K-edge at $\sim$ 7 keV and a broad peak around $\sim$ 30 keV, and the sum of the direct component (a power-law in 2 – 30 keV) and its reflection component can reproduce the observed spectra successfully [3]. A similar disk reflection feature is observed from numerous AGNs (see /35/ for a review), black hole candidates (e.g., /8, 11, 14, 54, 59/) and neutron star binaries (/60, 61/).

Ginga spectra in 2 – 25 keV are successfully fitted with a sum of the power-law function ($\alpha \sim 1.7$), a reflection component (slightly ionized, corresponding to $\xi_0 = 0.02$ in /58/) and a narrow iron emission line at 6.4 keV with the equivalent width $\sim$ 60 eV /14, 54, 55/. Note that the equivalent width is smaller than that expected from cold disk reflection (see below), and the line central energy is consistent with fluorescence from neutral or weakly ionized iron. Although energy resolution of *Ginga* (FWHM $\sim$ 20 %) is not as good as GSPC, evidence of the line red-shift and line broadening was not detected. Done et al. (1992) made a precise calculation of disk reflection spectra from ionized accretion disks, and applied their model to *EXOSAT* (GSPC and ME) and *HEAO1* archival data /62/. They found the disk is ionized such that FeXV is the most prominent species, and the iron line, having the equivalent width 44 ± 28 eV, is not required to be either redshifted or broadened. *BBXRT* observation of Cyg X-1 was made in 1990 /63/, with the first moderate energy resolution (FWHM $\sim$ 2.5 %) at the iron K-line band. The disk reflection feature was found, but the edge was at 7.1 keV, indicating Fe is neutral and no evidence of ionization. There was little evidence of Fe emission, with the possible exception of a narrow line at 6.4 keV with an equivalent with of 13 ± 11 eV.

Numerous theoretical works have focused on disk reflection spectra and associated iron fluorescent emission. Monte Carlo calculations are carried out for the cold disk for different geometries /64, 65/. It was found that iron line equivalent width is strongly dependent on the inclination angle, and also affected by other factors such as incident spectra and geometry. For a face-on disk, typically $\sim$ 150 eV equivalent-widths are expected. The line profile would have a two-horn structure corresponding to the blue- and red-shift due to the disk rotation /48/, and further distorted by Comptonization /64, 65/ and strong gravitational fields /66, 67, 68, 69/. Effects of disk ionization to the line emission is studied /70, 71/. It was found that expected iron line equivalent width is strongly dependent on the disk ionization state, widely ranging from 20 to 400 eV. For galactic black hole candidates, most Fe is assumed to be more ionized than FeXVII, and therefore resonance absorption and subsequent Auger process will suppress the iron line emission /70/. Furthermore, Comptonization in the hot, upper layer of the disk will broaden the emission line and make the line feature less perceptible. The broad K-shell absorption feature and weak emission line characteristic of Cyg X-1 may be thus explained /70/. However, in this case, rest-frame energy of the iron line from ionized disk in which FeXVII is dominant is $\sim$ 6.62 keV, as opposed to 6.4 keV.

Iron emission lines from most Seyfert I galaxies have the equivalent width $\sim$ 150 eV /35/, comparable to the theoretical prediction from cold accretion disks, so the cold disk reflection model seems to work primarily. The weak iron line feature from Cyg X-1 may be due to disk ionization /70/, and additionally due to Compton broadening in the upper layer of the disk /70/ and in the putative hot disk-corona /41/. If this is the case, the rest-frame central energy of the line will be $\sim$ 6.6 keV, and the line shape will be broadened and dramatically distorted due to the disk rotation, strong gravitational fields and Comptonization. Although current observations are consistent with a narrow line at 6.4 keV (this is rather unlikely from ionized accretion disks), the line expected from the ionized accretion disk will not be ruled out with the capability of contemporary instruments. To ascertain such a line feature, a superb energy resolution (to separate the line from the edge and determine its shape), wide energy coverage (to determine the continuum shape precisely) and high statistics are all required simultaneously.

---

high energies.

[3] If only 2 – 15 keV energy range is used, *Ginga* spectra can be fitted with a power-law + a broad emission line at $\sim$ 6.2 keV, the same spectral model used by Barr et al. (1985) /54/.

OTHER SPECTRAL TOPICS

Intensity dips

Irregular intensity dips are often seen near the superior-conjunction. The intensity drop is more prominent at lower energies, suggesting the dip is related to the photo-absorption due to intervening matter. A high quality dip spectrum was obtained with *Tenma* GSPC, and an iron K-edge was unveiled at 7.18 ± 0.18 keV. This indicates the absorption matter is neutral /72/, ruling out the speculation based on a lower energy observation that intervening material is ionized /73/. Comparison of the edge-depth and attenuation of the continuum above $\sim$ 4 keV suggests overabundance of iron by the factor $\sim$ 2 /72/. A low energy excess was seen during the dips below $\sim$ 4 keV, much greater than that expected from the amount of absorption above $\sim$ 4 keV /63, 72/. It was proposed that inter-stellar dust scattering may account for the soft-excess (/74/; see below), but variation of the soft-excess component during the dip /75/ supports the idea of partial absorption /72/, and only a small part of the soft-excess may be attributed to the dust scattering.

Dust scattering

Inter-stellar grains will scatter X-rays, and the scattered light will be observed as X-ray halos around point sources. Scattered photons travel longer path-lengths than direct ones, and thus should have remarkable time delays ($\sim$ 0.2 day for Cyg X-1). Scattered X-ray halos have been observed from several sources, and, in most analysis, halo surface brightness was derived as residual of fitting the source radial profile with telescope Point Spread Functions (PSF). Bode et al. (1985) studied the dust-scattering halo of Cyg X-1 by utilizing time variations of the source /76/. Photons detected at the core of the PSF show rapid time variations intrinsic to the source, but dust-scattered photons, detected at outer annuli, show little time variations being smeared out. The dust-scattering component, thus separated from the direct component, was detected up to $\sim$ 5' from the source, and its flux was $\sim$ 12 % of the total source flux. Because of the spectral incapability of the instrument (*Einstein* HRI), energy spectrum of the dust-scattering comonent could not be obtained.

Low energy spectral feature

Even outside of the dip, soft-excess emission below 2 – 3 keV, which cannot be accounted for by the power-law spectrum determined above 3 keV, has been reported from Cyg X-1 (e.g., /56, 77/), as well as from other black hole candidates /11, 54, 59/. Dust-scattering component will show a soft spectral feature ($\approx$ original spectrum times E^{-2}), but observed spectral variations of the soft component in moderately short timescales ($\sim$ an hour) (/11, 56/) suggest a different origin. Bałucińska and Hasinger (1991) /56/ found the dip spectra can be fitted with the two component model absorbed by cold matter, and the absorption column for the soft component is much smaller than that for the hard component. This suggests the soft component is extended as large as the size of the accretion disk, and only a part of the soft component is absorbed by cold material which originates in the stellar wind or in the outer parts of the accretion disk.

A broad emission-line like feature was found at 600 – 800 eV by the *EXOSAT* Transmission Grating Spectrometer /77/. This may be related to iron L-shell or ionized oxygen K-shell transitions. It should be noticed that disk reflection spectra in lower energies are enhanced by disk-ionization through reduction of the photo-electric cross section, and may yield some soft excess emission. Absorption edges and emission lines will be expected as well in 0.3 – 3 keV due to ionized heavy elements in the accretion disk /70/.

GINGA RE-ANALYSIS RESULTS

Ginga observed Cyg X-1 in 1987 August, 1990 May and 1991 June. In the latter two observations, the gain was reduced so that the energy range covers 2 – 60 keV, and the last one was carried out simultaneously with *GRO/OSSE*. Preliminary *Ginga* Cyg X-1 spectral results, using a disk reflection model, were presented at several occasions /14, 54, 55/. Although these results are almost correct, the reflection model used in these analysis is based on the calculation by Lightman

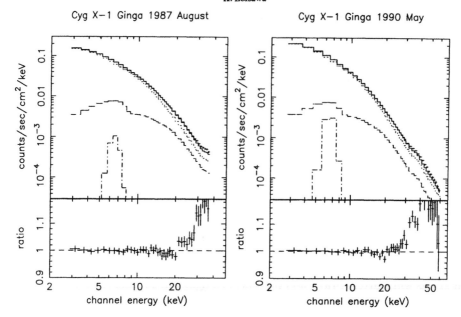

Figure 1: *Ginga* energy spectra of Cyg X-1 observed in 1987 August (left) and 1990 May (right) fitted with a power-law + disk reflection model by Done et al. (1992) /62/. The fitting was carried out in 3 – 20 keV, and data in higher energy bins are shown together. It is seen that the model fit is successful in the adopted energy range, and there is an excess above ∼ 25 keV.

and White (1988) for fixed photon-index and ionization state /58/, and hence may not be correct for different photon-indices and ionization states. Furthermore, Cyg X-1 is bright and has a hard spectrum; it is even brighter than the Crab nebula, a calibration source, at the highest LAC energy range. Consequently, systematic uncertainties, as opposed to statistical errors, make the data analysis somewhat laborious and difficult. For example, the standard particle background monitor, which is the count-rates over the upper PHA discriminator, is not usable being contaminated by the hard X-rays from the source. In addition, the LAC collimator reflects some soft X-rays and changes continuum shape slightly, and may affect subtle iron line features.

We have been working carefully to eliminate/estimate these systematic effects (see /59/ for the collimator reflection correction), and finally obtained reliable spectra and responses. We adopted the reflection model by Done et al. (1992) /62/, which was successfully applied to *HEAO1/EXOSAT* Cyg X-1/62/, and *Ginga* GX339–4/59/ and X1608–522/61/ energy spectra. In this model, ionization state of the disk is described by the ionization parameter ξ ($\equiv L/nr^2$), and directly constrained from model fitting. In order to avoid effects of the soft-excess component, we truncated the data below ∼ 3 keV. In figure 1, we show results of model fitting to typical energy spectra outside of the dip in 1987 and 1990. The model is a power-law + disk reflection + a narrow iron-emission line (fixed at 6.4 keV) with low energy absorption (fixed to the inter-stellar value $N_H=6\times10^{21}\mathrm{cm}^{-2}$), and fitting is carried out in the energy range 3 – 20 keV. We found the model fit is successful in this energy band, but there is a 10 – 15 % excess above the model in the higher energy range (this will be discussed below). Important spectral parameters, for 1987 and 1990 data respectively, are the following; photon-index = 1.73 and 1.76, normalization of the reflection component relative to the direct one (= $\Omega/2\pi$) = 0.72 and 0.51, ionization-parameter ξ (disk temperature is fixed to 10^5 K) = 93 and 124, and equivalent-width of the line = 19 and 64 eV. Moderately ionized disks are favored, but the neutral disk ($\xi = 0$) is also acceptable. It should be noted that iron-line parameters (central energy, intrinsic width and intensity) are hardly constrained by the data. The model without the iron-line is perfectly acceptable (the *F*-test does not confirm the existence of the iron line), and on the other hand, a broad, strong iron-line is also consistent with the data (the upper-limit of the line-width is not constrained). For a broad-line at 6.4 keV having a 1 σ width of 0.5 keV, the 90 % upper-limit of the equivalent width is $\sim$ 150 eV, and the line intensity can become still larger for broader lines. The iron line and edge structures cannot be separated, and those parameters are coupled to each other. For example, the ξ of the disk tends to be smaller for stronger lines; this is because the iron-edge depth is deeper for ionized disks, so a strong line is needed for the neutral disk to explain the observed edge depth.

The hard-excess at $E \gtrsim 25$ keV was unexpected, since continuum emission is believed to be from thermal Comptonization, and anticipated, on the contrary, to roll-over at high energies. The fact that the hard-excess feature is seen from almost all the spectra, regardless of the observation date and gain-setting, reinforces its reality. Recently, a new calculation was made by Magdziarz and Zdziarski (1995) for disk reflection spectra taking account of viewing angle dependencies /78/. According to their calculation, the reflected spectrum strongly hardens with increasing the viewing angle (approaching to face-on), and this effect is most significant in hard X-rays and soft γ-rays. Cyg X-1 will have the inclination angle $\sim 30°$/79/, and thereby the reflection spectrum is expected to be $\sim$ 50 % stronger at $\sim$ 30 keV than that expected from the angle-averaged reflection model, on which Done et al. (1992)'s calculation basically stands. The observed hard-excess in the *Ginga* spectra may thus be explained. There are *Ginga* and *GRO/OSSE* simultaneous Cyg X-1 (1991 June) and GX339-4 (1991 September) observations, and the project to study hard X-ray spectral features from these data and test the reflection model by Magdziarz and Zdziarski (1995) is now going. Future *XTE* observations will be expected to clarify the hard X-ray spectral feature in detail.

ASCA PRELIMINERY RESULTS

The *ASCA* satellite was launched in February 1993/80/, carrying the two focal plane imaging instruments, SIS (X-ray CCDs) and GIS (position sensitive GSPC). The distinguishing character of *ASCA* is superb spectral capability up to $\sim$ 10 keV; the energy resolution at 6 keV is $\sim$ 2 % and $\sim$ 8 % (FWHM) for SIS and GIS respectively. *ASCA* observed Cyg X-1 twice in the performance verification phase, in October and November 1993, and again in November 1994 to calibrate the telescope point spread function. Although the SIS in principle has a potential to reveal iron line features with an unprecedented energy resolution, unfortunately, Cyg X-1 is too bright for the SIS to obtain reliable spectra. For bright sources, two or more X-ray photons may hit adjacent CCD pixels within a single exposure (typically 4 sec), and subsequently are recognized as a single false event (*double event* problem). At least at the end of 1994, the SIS has never been calibrated for such bright sources as Cyg X-1, and there is no way to correct the double event problem. Hence, below, I will mention only GIS spectral results of Cyg X-1. The result presented here is essentially the same as that in /4/.

In November 1993, Cyg X-1 was observed for $\sim$ 50 ksec with the standard mode that is suitable for precise spectral study. A spectral feature was seen at around Xe-LIII edge (4.78 keV), but this is probably instrumental, because Cyg X-1 is so bright and current XRT/GIS calibration uncertainties would be much larger than the statistical errors. In order to investigate iron K-structure, avoiding the Xe-L feature and effects of the soft-excess component, the data in 5.5 – 10 keV range were fitted with a power-law function (figure 2; left). Clearly, the proclaimed broad-line like feature and iron edge structure are seen. Next, we applied a disk reflection model by Done et al. (1992) /62/. The model was formally acceptable ($\chi^2 = 94.8$ for 92 d.o.f.), but the residual suggests existence of an iron emission line (figure 2; right). By introducing a narrow line at 6.4 keV, the χ^2 is reduced to 71.9 for 91 d.o.f. (F-value $= (94.8\text{-}71.9)/(71.9/91) = 29$), indicating the confidence of the line is more than 99.9999 %. Important parameters for the power-law + reflection + narrow line model are the following; $\alpha = 1.90$, $\Omega/2\pi = 1.056$, $\xi = 17$ (T = fixed at 10^5 K), and line equivalent-width = 13 eV. The intrinsic line width is constrained to $\lesssim 0.1$ keV (90 %). The energy and narrowness of the line suggests its origin being from whether surface of the companion star or outer parts of the accretion disk. The inner, relativistic region of the disk is unlikely to emit the narrow 6.4 keV line.

K. Ebisawa

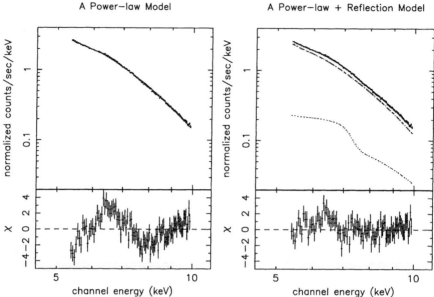

Figure 2: *ASCA* GIS energy spectra of Cyg X-1 observed in November 1994. Left: fit with a power-law function ($\alpha = 1.83$). Right: the power-law + reflection spectra for the best-fit power-law + reflection + narrow line model. So that the line feature is visible in the residual, the line model was taken away from the best-fit model.

Although this model can fit the spectrum in the 5.5 – 10 keV band well, the steep power-law slope (1.90) does not seem consistent with the low energy spectra. If this slope is extended down to below ~ 1 keV, the absorption column needs to be nearly 10^{23} cm^{-2}, but this is very different from the commonly accepted value, $(6 - 7) \times 10^{21}$ cm^{-2}. By using the entire GIS energy band, and also from *Ginga* and other observations having wide energy coverage, photon index values of 1.7 – 1.8 seem more plausible. However, if such flatter slopes are assumed, the reflection + narrow line model does not fit the 5.5 – 10 keV data any more. This problem has yet to be solved, but one possibility may be that there is a broad line centering at ~ 6.2 keV as well as the narrow line and reflection component. If this is the case, a part of the flux below the iron edge is ascribed to the broad line, and thus the photon-index may be remedied to 1.7 – 1.8. In order to have a consistent model to explain the entire GIS spectrum, detailed GIS data analysis is in progress, as well as finalizing the XRT/GIS calibration.

SUMMARY

In this paper, I have reviewed X-ray spectral characteristics of Cyg X-1 on the published material, *Ginga* re-analysis results, and *ASCA* preliminary results. One of the major observational difficulties in studying Cyg X-1 (and other galactic black hole candidates) is, as we saw, to determine the continuum spectral shape and the subtle iron line/edge structure individually, which are often coupled and hard to seperate. In order to do that, we need both high energy resolution (e.g., 2 % with CCD) and wide energy coverage (e.g., 2 – 60 keV with proportional counters) simultaneously, but they are not compatible in any single detectors. In this regard, simultaneous observations with detectors of complementary capabilities will be encouraging. Simultaneous *ASCA*, *XTE* and *GRO* observations in near future will be useful, and in further future, observations by grating spectrometers on *XMM/AXAF* and any hard X-ray/γ-ray missions are expected to provide new insights into the nature of Cyg X-1.

ACKNOWLEGEMENT

I would like to acknowledge all the members of the *Ginga* and *ASCA* teams for making the Cyg X-1 observations available. The *Ginga* Cyg X-1 work presented here is a collaboration with Dr. C. Done and Mr. Ueda. I am grateful to Mr. Ueda for helping the *ASCA* Cyg X-1 data analysis. I would like to thank Dr. C. S. R. Day, the Scientific Editor of this session, for his continuous encouragement and carefully reading the manuscript.

REFERENCES

1. Paczyński, S. 1974, Astron. and Astrophys. 34, 161

2. Oda, M. 1977, Space Sci. Review, 20, 757

3. Liang, E. P. and Nolan, P. L. 1984, Space Sci. Rev. 38, 353

4. Tanaka, Y. 1994, in *New Horizon of X-ray Astronomy*, ed. F. Makino and T. Ohashi, p. 37, Universal Academy Press, Tokyo

5. Cowley, A. P. 1994, in *The Evolution of X-ray Binaries*, ed. S. S. Hold and C. S. Day, p. 45, American Institute of Physics, New York

6. White, N. E. 1994, p. 53, in the same volume as /5/

7. Cowley, A. P. 1992, Ann. Rev. Astron. Astrophys., 30, 287

8. Inoue, H. 1992, in *Frontiers of X-ray Astronomy*, ed. Y. Tanaka and K. Koyama, p. 291, Universal Academy Press, Tokyo

9. Ricketts, M. 1983, Astron. and Astrophys., 118, L3

10. Kitamoto, S. et al. 1992, ApJ, 394, 609

11. Ebisawa, K. et al. 1994, PASJ, 46, 375

12. Sanford, P. W. et al. 1975, Nature, 256, 109

13. Tanaka, Y. 1989, in Proceedings of the 23rd ESLAB Symposium, vol. 1., p. 3, ESA, Noordwijk, the Netherlands

14. Ebisawa, K. 1991, Doctoral Thesis, Univ. of Tokyo (ISAS Research Note 483)

15. Ebisawa, K. et al. 1993, ApJ, 403, 684

16. Mineshige, S. et al. 1994, ApJ, 426, 308

17. Ebisawa, K. 1994, p. 143, in the same volume as /5/

18. Shimura, T. and Takahara, F. 1995, ApJ, in print

19. Grebenev, S. A. et al. 1994, p. 61 in the same volume as /5/

20. Ballet, J. et al. 1994, p. 131 in the same volume as /5/

21. Miyamoto, S. et al. 1991, ApJ, 383, 784

22. Ichimaru, S. 1977, ApJ, 214, 840

23. Inoue, H. and Hoshi, R. 1987, ApJ, 322, 320

24. Hoshi, R. and Inoue, H. 1988, PASJ, 40, 421

25. Tanaka, Y. 1994, IAU circ. 5877

26. Sunyaev, R. A. and Trümper, J. 1979, Nature, 279, 506

27. Sunyaev, R. A. and Titarchuk, L. G. 1980, Astron. and Astrophys., 86, 121

28. Salotti, L. et al. 1992, Astron. and Astrophys., 253, 145

29. Grabelsky, D. A. et al. 1992, in *Compton Gammra-ray Observatory*, ed. M. Friedlander, N. Gehrels and D. J. Macomb, p. 345, American Institute of Physics, New York

30. Apparao, K. M. V. 1984, Astron. and Astrophys., 375, 139

31. Shapiro, S. L., Lightman, A. P. and Eardley, D. M. ApJ, 204, 187

32. Liang, E. P. T. and Price, R. H. 1977, ApJ, 218, 247

33. Bisnovatyi-Kogan, G. S. and Blinnikov, S. I. 1977, Astron. and Astrophys., 59, 111

34. Galeev, A. A., Rosner, R. and Vaiana, G. S. 1979, ApJ, 229, 318

35. Mushotzky, R. F., Done, C. and Pounds, K. A. 1993, Ann. Rev. Astron. Astrophys., 31, 717

36. Titarchuk, L. 1994, ApJ, 434, 570

37. Ling, et al. 1987, ApJ, 321, L117

38. Liang, E. and Dermer, C. D. 1988, ApJ, 325, L39

39. Ling, J. C. and Wheaton, W. A. 1989, 343, L57

40. Gilfanov, M. et al. 1994, ApJS, 92, 411

41. Haardt, F. et al. 1993, ApJ, 411, L95

42. Haardt, F. and Maraschi, L. 1993, ApJ, 413, 507

43. Haardt, F. 1993, ApJ, 413, 680

44. Titarchuk, L. 1994, ApJ, 434, 570

45. Titarchuk, L. And Mastichiadis, A. 1994, ApJ, 433, L33

46. Hua, X. And Titarchuk, L. 1995, ApJ, in print

47. Barr, P., White, N.E., and Page, C.G. 1985, MNRAS, 216, 65p

48. Fabian, A. C. et al. 1989, MNRAS, 238, 729

49. Stella, L. 1990, Nature, 344, 747

50. Kitamoto, S., Takahashi, K., Yamashita, K., Tanaka, Y., and Nagase, F. 1990, PASJ, 42, 85

51. Makino, F. and Astro-C team 1987, Astrophys. Lett. Comm. 25, 223

52. Turner, M. J. L. et al. 1989, PASJ, 41, 345

53. Inoue, H. 1989, vol 2, p. 783, in the same volume as /13/

54. Tanaka, Y. 1992, in *Iron Line Diagnostics in X-ray Astronomy*, ed. A. Treves, G. C. Perola and L. Stella, p. 80, Springer-Verlag, Berlin

55. Ebisawa, K. et al. 1992, p. 301, in the same volume as /8/

56. Bałucińska, M., Assignor, G. 1991, Astron. and Astrophys. 241, 439

57. Guilbert, P. W. and Rees, M. J. 1988, MNRAS, 233, 475

58. Lightman, A. P. and White, T. R. 1988, ApJ, 335, 57

59. Ueda, K., et al. 1994, PASJ, 46, 107

60. Mitsuda, K. 1992, p. 115 in the same volume as /8/

61. Yoshida, K., et al. 1993, PASJ, 45, 605

62. Done, C. et al. 1992, ApJ, 395, 275

63. Marshall, F.E., Mushotzky, R. F., Petre, R., and Serlemitsos, P. J. 1993, ApJ, 419, 301

64. Matt, G. Perola, G. C. and Piro, L. 1991, Astron. and Astrophys., 247, 25

65. George, I. M. and Fabian, A. C. 1991, MNRAS, 249, 352

66. Kojima, Y. 1991, MNRAS, 250. 629

67. Laor, A. 1991, ApJ, 376, 90

68. Matt, G., Perola, G. C., Piro, L., and Stella, L. 1992, Astron. and Astrophys., 257, 63

69. Matt, G., Perola, G. C., and Stella, L. 1993, Astron. and Astrophys., 267, 643

70. Ross, R. R. and Fabian, A. C. 1993 MNRAS, 261, 74

71. Matt, G., Fabian, A. C., and Ross, R. R. 1993, MNRAS, 262, 179

72. Kitamoto, S. et al. 1984, PASJ, 36, 731

73. Pravdo, S. H. et al. 1980, ApJ, 237, L71

74. Xu, Y., McCray, R. and Kelley, R. 1986, Nature, 319, 652

75. Kitamoto, S. Miyamoto, S. and Yamamoto, T. 1989, PASJ, 41, 81

76. Bode, M. F. et al. 1985, ApJ, 299, 845

77. Barr, P. and van der Woerd, H. 1990, ApJ, 352, L41

78. Magdziarz, P. and Zdziarski, A. A. 1995, MNRAS, in print

79. Gies, D. R. and Bolton, T. 1986, ApJ, 304, 371

80. Tanaka, Y., Inoue, H. and Holt, S. S. 1994, PASJ, 46, L37

Pergamon

Adv. Space Res. Vol. 19, No. 1, pp. (1)15–(1)23, 1997
© 1997. Published by Elsevier Science Ltd on behalf of COSPAR
Printed in Great Britain. All rights reserved
0273–1177/97 $17.00 + 0.00

PII: S0273-1177(97)00031-8

SPECTRAL STATES OF GALACTIC BLACK HOLE CANDIDATES: RESULTS OF OBSERVATIONS WITH ART-P/GRANAT

S. A. Grebenev, R. A. Sunyaev and M. N. Pavlinsky

Space Research Institute, Russian Academy of Sciences, Profsoyuznaya 84/32, Moscow 117810, Russia

ABSTRACT

Observations of Galactic black hole candidates performed with the ART-P telescope aboard *Granat* in 1990-1992 focused our attention on the problem of studying spectral variability of these sources as a function of the accretion rate. We show that in contrast with the current models of disk accretion onto a black hole the hard spectral state of the sources in which they exhibit a Cyg X-1 type hard X-ray spectrum always occurs at smaller accretion rate than the soft state. The arguments that hard X-rays detected during the "very high" and "low" states of black hole candidates have a distinct origin are presented.

INTRODUCTION

Two emission components (soft and hard) observed in X-ray/γ-ray spectra of many stellar mass black hole candidates are usually explained in terms of two physically distinct regimes of disk accretion onto a black hole /1/. The soft component bright at the energies $h\nu \leq 8$ keV is produced in the outer optically opaque disk region. Its spectrum is successfully described with a multicolor disk black-body model. It is assumed that this component gives a dominant contribution to the source overall luminosity only at small accretion rates ($\dot{m} = \dot{M}/\dot{M}_{ed} \ll 1$, where $\dot{M}_{ed}$ is the Eddington critical accretion rate). On the contrary, the hard X-ray emission originates in the hot ($kT_e \sim 50$ keV) optically thin plasma of the innermost region. The observed spectrum extended up to $h\nu \sim 200$ keV is formed due to the inverse Compton scattering of low energy photons by high temperature electrons /2, 3/. In the case of accretion onto a Schwarzschild black hole, the hard component appears in the spectrum if the accretion rate $\dot{m}$ exceeds $1/110\,(\alpha m_*)^{-1/8}$ /1/. Here $m_* = M/10\,M_\odot$, M is the black hole mass, α - the viscosity parameter.

Long-term observations of several Galactic black hole candidates performed with the ART-P telescope aboard *Granat* in 1990-1992 provided an opportunity for studying spectral variability of the sources vs. the accretion rate. In particular, X-ray luminosity of GX339-4 was measured during its different states. The drop by a factor of 10 in X-ray brightness of 1E1740.7-2942 was discovered. However the most spectacular was the behaviour of Nova Muscae 1991 (GRS1124-68). After the outburst the source exhibited step-by-step three distinct spectral states (two-component, soft and hard) while its overall luminosity declined gradually. It was a great surprise for us that in contrast with the existing accretion disk models the hard state of these sources always occurred at smaller accretion rates than the soft one.

X-RAY OBSERVATIONS OF BLACK HOLE CANDIDATES

Bimodal X-ray behaviour was first discovered in Cyg X-1, a most reliable and best studied black hole candidate. More than 90% of the time the source spends in a "low" state of intensity exhibiting a hard power-law spectrum with the photon index $n \simeq 1.6$ /4/. During a transient "high" state the flux in the 2-10 keV band increases by a factor of 5 and the spectrum steepens to $n \simeq 4$ /5/. It is difficult to say in which of the two states the source overall luminosity is higher. The low-state

GRANAT/ART-P/SIGMA

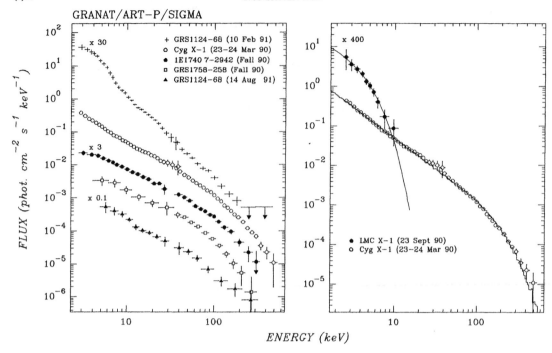

Fig. 1. The hard-state spectra of several black hole candidates (left panel) and the soft-state spectrum of LMC X-1 (right panel) obtained with the ART-P and SIGMA instruments aboard *Granat* /6/ and given in comparison with the spectrum of Cyg X-1, the most reliable black hole candidate.

spectrum integration shows that the hard X-ray band contains a major part of the luminosity, $\sim 3 \times 10^{37} \, (d/2.5 \text{ kpc})^2$ erg s^{-1} where d is the distance to the source (see e.g. /6/ for the results of the luminosity measurements with *Granat* in the 3-200 keV energy band). This is roughly 50 times lower than the critical Eddington luminosity for a 10 $M_\odot$ black hole ($L_{ed} \simeq 1.3 \times 10^{39} \, m_*$ erg s^{-1}), i.e. $\dot{m} \lesssim 0.02 \, m_*^{-1}$. In the high state the contribution of the soft $h\nu \leq 2$ keV X-ray and EUV spectral bands becomes significant, so the determination of the bolometric luminosity is complicated. As a rule high-state observations of Cyg X-1 were incomplete and performed in the narrow 2-10 keV band. Recently two other X-ray sources were suspected to be good potential black hole candidates /7/. Those are 1E1740.7-2942, the hardest source in the Galactic center region, and GRS1758-258, a source discovered with *Granat* in the vicinity of GX 5-1. Both sources have a Comptonized Cyg X-1 type spectrum (see left panel in Figure 1) slightly flatter in the standard X-ray band ($n \simeq 1.4$). Their 3-200 keV luminosity is normally also very close to that of Cyg X-1, 2.7 and $1.9 \times 10^{37} \, (d/8.5 \text{ kpc})^2$ erg s^{-1} for 1E1740.7-2942 and GRS1758-258 respectively /6/. Note that no significant spectral change was detected during the drop of 1E1740.7-2942 intensity occurred in the beginning of 1991. The source remained in the hard state in spite of the total luminosity decrease by a factor of 10 /8, 9/.

Unlike these Galactic sources, LMC X-1 and LMC X-3, two black hole candidates located in the Large Magellanic Cloud, spend most of the time in the soft state. In Figure 1 (right panel) the typical photon spectrum of LMC X-1 obtained with ART-P in the fall of 1990 is presented in comparison with that of Cyg X-1. It was scaled for clarity to the 2.5 kpc distance of Cyg X-1 (by 400 times !). According to White & Marshall /10/ the luminosities of both sources are at the level of $\sim 2 \times 10^{38} \, (d/50 \text{ kpc})^2$ erg s^{-1} in the 0.5-60 keV band which corresponds to the accretion rate $\dot{m} \simeq 0.2 \, m_*^{-1}$ one order of magnitude higher than that typical for the hard-state sources.

Due to the dispersion in $\dot{M}_{ed}$ values of individual sources caused by the obvious variety in black hole masses the observation of spectral transitions from the same source should be most indicative (from the point of view of the spectral hardness – accretion rate analysis). GX339-4 provides us with such a possibility. In 1990-1992 all the three known spectral states of the source (soft, hard

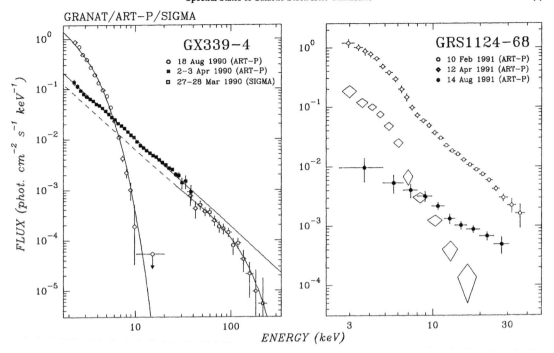

Fig. 2. Photon spectra of GX339-4 and Nova Muscae (GRS1124-68) measured with *Granat* during the various states of the sources

and off) were investigated with ART-P /6, 11/. The spectra measured during these states are given in the left panel of Figure 2. The source luminosity observed above 3 keV was equal to 0.5, 1.9 and 0.07×10^{37} $(d/4 \text{ kpc})^2$ erg s^{-1} in the soft, hard and off (3-σ upper limit) states respectively. It is important to emphasize that during the soft state only a small portion of the disk luminosity was detectable in rather a hard 3-35 keV energy band of ART-P. Approximation of the GX339-4 spectrum with the multicolor disk black-body model allows its overall luminosity to be evaluated using the best-fit parameters, the maximum black-body temperature of the disk, kT_{max}, and the radius where it is reached, r_{max}. Such an estimate, $L_B \simeq 40\, r_{max}^2 \sigma T_{max}^4 \simeq 5.5 \times 10^{37} \cos(i/60°)^{-1}$ erg s^{-1} /11/ where i is the disk inclination and σ is the Stefan-Boltzmann constant, shows that more than 90% of the disk soft-state luminosity is emitted in the soft $h\nu \leq 3$ keV X-ray and EUV bands. The measurements performed with *Ginga* in 1988 during the "very high" state of GX339-4 demonstrated that its 1-37 keV luminosity could reach 6×10^{37} erg s^{-1} /12/. In that very high state the source had a two-component spectrum consisting of a bright soft component and a hard power-law tail.

BRIGHT X-RAY NOVAE

The observations carried out with X-ray instruments aboard *Mir-Kvant*, *Ginga*, *Granat* and *Compton Observatory* in 1987-1993 indicated that bright X-ray novae flared up in the few kpc vicinity of the Earth approximately once a year. These sources are likely to form the most numerous class of the Galactic binaries containing black holes /13/. Studying of X-ray novae in quiescent when the accretion disk contributes nothing to the optical light allowed the mass functions for the most of them to be determined. The mass lower limits of the compact objects in A0620-00 /14/, GS2023+338 /15/ and GRS1124-68 /16/ were estimated to be greater than $3M_\odot$, the maximum mass of a neutron star.

X-Ray Nova Muscae 1991 (GRS1124-68) was extensively studied with ART-P during nine months after the outburst /13/. The observed spectra can be divided up into three groups distinctive by both the spectral type and observing epoch (the right panel of Figure 2). The spectrum was either two-component with a bright disk black-body component and a steep ($n \simeq 2.5$) hard power-law tail (days 8-33 after the outburst), or soft, similar to the high-state spectrum of GX339-4 (day 94 - the only *Granat* observation occurred that time), or hard, characterized by $n \sim 1.6$ and resembling the low-state spectra of Cyg X-1 or GX339-4 (days 143-219). The described spectral variability reproduced mainly irregular changes in the hard emission component. The soft 3-8 keV flux decreased exponentially with the e-folding time of about 35-40 days during the whole time of observations with the exception of the day ~ 70 when a sudden flux increase by 2.3 times ("kick" in the light curve) was detected with *Ginga* /17/. The outbursts of X-ray novae are believed to be connected with episodic enhancements of the disk accretion rate onto a black hole in binary systems. It is natural to assume that the monotonous decay of the soft X-ray light curve of Nova Muscae (and the observed succession of spectral transitions) reflects a gradual decrease of $\dot{m}$. The 3-200 keV luminosity of GRS1124-68 was equal to 3.7, 0.23 and 0.26×10^{37} $(d/3 \text{ kpc})^2$ erg s^{-1} in the very high, high and low states respectively /13, 18/. For the luminosity in the off-state (observed after the source switch-off) the 3 σ upper limit, 8×10^{35} $(d/3 \text{ kpc})^2$ erg s^{-1}, was obtained. According to *Ginga* performing measurements in the softer $h\nu \geq 1$ keV band the source luminosity was at the level of 30 and 6×10^{37} $(d/3 \text{ kpc})^2$ erg s^{-1} nearly about the dates when the very high and high states were detected with ART-P /17/.

There were another four X-ray novae observed, A0620-00 (Nova Monocerotis 1975), H1705-25 (Nova Ophiuchi 1977), GS2000+25 (Nova Vulpeculae 1988) and GRS1009-45 (Nova Velorum 1993) /19-22/, which exhibited X-ray properties common with those of Nova Muscae 1991. We mean such properties as an exponential decline of the light curve, the ultrasoft X-ray spectrum with a power-law tail present during the first weaks after the outburst, the hard Cyg X-1 type spectrum – few months later, etc. The behaviour of four other known long-lived X-ray transients, GS2023+338 (Nova Cygni 1989), GRO J0422+32 (Nova Persei 1992), GRS1716-249 (Nova Ophiuchi 1993) and probably GRO J1655-40 (Nova Scorpii 1994), was quite different /20, 22-24/. These sources could be designated as "hard" X-ray novae because there was no distinct soft X-ray component found in their spectra. At least for some of these sources the absence of the soft spectral component can be attributed to rather a low $\dot{m} \lesssim 0.02\, m_*^{-1}$ accretion rate at the peak of their X-ray brightness. As follows from the observations of black hole candidates, at such a low accretion rate the sources should be in the hard spectral state and have the Comptonized X-ray spectrum. On the other hand note that the observed behaviour of "hard" X-ray novae was so strange and complex (e.g. the peculiar behaviour of Nova Cygni 1989 which was violently variable in X-rays and demonstrared the complex effects of absorption in its spectrum /20/ or the discovery of twin-jet ejection of relativistic plasma from the central object in Nova Scorpii 1994 /25/) that lets us to suggest the accretion rate to be not the unique parameter to determine their properties. In the case of "soft" X-ray novae, the accretion rate at the peak of their X-ray brightness was much higher and close to the critical Eddington limit, so the two-component spectra were observed.

DISK LUMINOSITY IN THE SOFT X-RAY AND EUV BANDS

The problem with the portion of the accretion disk luminosity emitted in the soft X-ray and EUV bands requires additional discussion. In Figure 3 (top panel) we present the total and $h\nu \geq 3$ keV disk luminosities computed as functions of the photon flux measured at 3 keV (in the softest ART-P channel) at a distance of 1 kpc. The computations are based on the multicolor disk black-body approximation /1/ to the spectrum. Two assumptions for the mass of a compact object are considered, $M = 10\, M_\odot$ (solid lines) and $3\, M_\odot$ (dotted lines). The figure obviously demonstrates that during the very high and high states of black hole candidates when the bright disk black-body component is present in their spectra the soft $h\nu \leq 3$ keV band contains a major ($\geq 90\%$) portion of the total disk luminosity*. The contribution of the soft X-ray and EUV spectral bands to the source hard-state luminosity can be evaluated in the same manner. Let us assume that low-energy

---

*As was first mentioned by Shakura & Sunyaev /1/, multiple Thomson scattering of photons in the opaque disk modifies the disk emission spectrum increasing its brightness at high energies. This increase can be one of the reasons why the results of luminosity measurements with ART-P in the 3-30 keV band during the soft states of GX339-4 and some other black hole candidates (see crosses in the same figure) lie slightly above the predicted curve. The distinction between disk inclination angles of individual sources and that adopted in our computations, $i = 60°$, can be another reason.

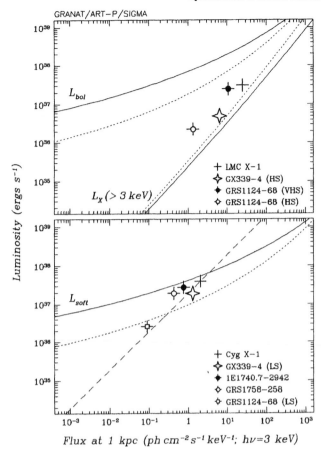

GRANAT/ART−P/SIGMA

L_{bol}

$L_X (> 3\ keV)$

+ LMC X−1
◇ GX339−4 (HS)
◆ GRS1124−68 (VHS)
◇ GRS1124−68 (HS)

L_{soft}

+ Cyg X−1
◇ GX339−4 (LS)
◆ 1E1740.7−2942
◇ GRS1758−258
◇ GRS1124−68 (LS)

Luminosity (ergs s⁻¹)

Flux at 1 kpc (ph cm⁻² s⁻¹ keV⁻¹; hν=3 keV)

Fig.3. Disk black-body luminosity as a function of photon flux at 3 keV (for a distance of 1 kpc and disk inclination $i = 60°$, solid and dotted lines correspond to the accretion onto $10\ M_{\odot}$ and $3\ M_{\odot}$ black holes, respectively). In the top panel, the overall disk luminosity is compared with that emitted above 3 keV. The results of measurements with ART-P in the 3-30 keV band during the very high (VHS) and high (HS) states of several black hole candidates are shown after the flux reduction to the 1 kpc distance. In the bottom panel, we present the upper limit for the soft X-ray and EUV luminosity of black hole candidates during their low/hard states (LS). It was suggested that 1/4 of the flux, measured from these sources at 3 keV, could be attributed to the disk black-body component. The hard ($h\nu \geq 3$ keV) luminosity observed with *Granat* during the low states of several sources is presented. The dashed line shows the hard luminosity dependence on the 3 keV flux for a source with a Cyg X-1 type spectrum.

photons with the disk black-body spectrum are responsible for 1/4 (the typical 3σ upper limit) of the photon flux at 3 keV measured with ART-P during the observations of black hole candidates in the low/hard states. The corresponding upper limit for the luminosity in the soft disk black-body spectral component is presented in the bottom panel of Figure 3 by solid and dotted lines (for the accretion onto $10 M_{\odot}$ and $3 M_{\odot}$ black holes respectively). For comparison, the hard $h\nu \geq 3$ keV luminosity measured with *Granat* during the observations of several sources in their low states is shown by crosses. The figure argues that the disk low-state luminosity in the soft X-ray and EUV bands can not significantly exceed that emitted in the high energy band.

TWO DIFFERENT TYPES OF HARD X-RAY SPECTRA ?

The observations show that Galactic black hole candidates are strong emitters of hard $h\nu \geq 10$ keV X-rays only when they are in one of the two states of intensity: very high state or low state. In the very high state the source spectra are two-component with a bright soft disk black-body component and a power-law ($n \sim 2.5$) tail. In the low state hard Cyg X-1 type spectra are observed. They are also power-law but flatter ($n \sim 1.6$) in the $h\nu \lesssim 60$ keV X-ray band and are exponentially declining at higher energies. In both cases hard X-rays are most likely produced due to Comptonization of low-energy emission in the high temperature electron plasma. Thus, the observed spectral differences should reflect the differences in the plasma cloud parameters typical for that regime of disk accretion which corresponds to the given state.

In Figure 4 (left panel) we present the results of Monte-Carlo computations of hard X-ray spectra formed due to Comptonization of soft $h\nu_0 = 100$ eV photons in the disk cloud with electron temperature $kT_e = 57$ keV. Two shown spectra are obtained under assumption of different optical half-thickness of the disk with respect to Thomson scattering $\tau \simeq 1.0$ and 0.3. The first spectrum reproduces that of Cyg X-1 /3, 6/, the second – the hard X-ray tail present in the spectrum of

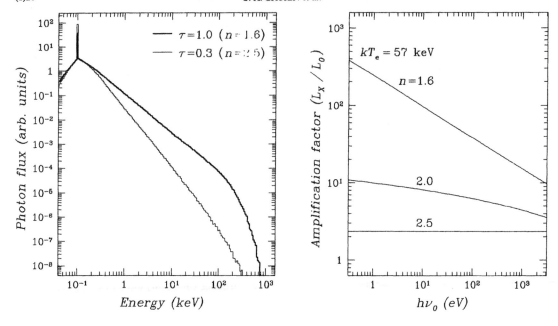

Fig. 4. Photon spectra resulting from Comptonization of low-energy $h\nu = 100$ eV photons in the high temperature $kT_e = 57$ keV plasma cloud. Varying the cloud optical depth we can reproduce the hard X-ray spectra of black hole candidates both in low ($n \sim 1.6$) and very high ($n \sim 2.5$) intensity states.

Fig. 5. Efficiency of Comptonization (the ratio of the hot plasma cloud luminosity to that of the embedded source of low-frequency photons as a function of the photon energy /26/. The curves correspond to different values of the photon index n of X-ray emission formed in the cloud. The changes in n reflect the changes in the cloud optical depth).

black hole candidates during their very high state. One of the most interesting features of the latter spectrum consists of the absence of the high energy cut-off similar to that visible in the Cyg X-1 type spectrum. In reality, the cut-off still exists but it is less obvious because the whole source spectrum is much steeper. It is well known that no high-energy cut-off was detected in the broad band X-ray spectrum of Nova Muscae 1991 (see Figure 1 or /18/ for more details). We show here that this result is fully in the course of the Comptonization model.

The observation of a steep ($n \sim 2.5$) hard X-ray spectrum in the very high state of Nova Muscae 1991 lets us to suggest that in this case the luminosity of the low-frequency photon source L_0 was amplified by Compton scattering on electrons in the high temperature region of the disk only to a small extent ($L_X/L_0 \simeq 2.5$, see /26/). Comparison of the luminosities measured in the soft and hard X-ray bands /13, 17/ shows that only rather a small ($\sim$ a few %) portion of the soft component photons could be reprocessed to the source hard X-ray radiation. This is a direct evidence that soft and hard components in the source spectrum are produced in geometrically separated parts of the accretion disk*. In the case of the Cyg X-1 type spectrum the luminosity amplification factor is higher one or two orders of magnitude (see Figure 5), so the luminosity of the source of low-frequency photons scattered in the plasma cloud is really negligible.

---

*Hard X-rays could also be produced in a hot corona the existence of which above the cool disk surface was assumed by a number of authors /27, 28/. In this case the portion of soft photons emitted by the disk and then scattered in the hot electron plasma of the corona would be of about its optical depth with respect to Thomson scattering ($\sim \tau \gtrsim 0.3$ as follows from our Monte-Carlo computations). Thus, the source luminosity in the hard X-ray band would be much higher than that actually observed with the ART-P and SIGMA instruments from Nova Muscae 1991.

ACCRETION DISK MODELS VS. OBSERVATIONS

The results of observations discussed above can be summarized as follows: *Depending on $\dot{m}$ stellar mass black hole candidates spend most of the time in one of the following four spectral states: very high state occurring at $0.2 \lesssim \dot{m} \lesssim 1$ and characterized by the two-component spectrum, high state ($0.02 \lesssim \dot{m} \lesssim 0.2$, ultrasoft spectrum), low state ($0.001 \lesssim \dot{m} \lesssim 0.02$, hard Comptonized spectrum) or off-state ($\dot{m} \lesssim 0.001$, unknown spectrum).* Thus, the hard state of the sources occurs at smaller accretion rates than the soft one.

It is typically assumed that the hard Cyg X-1 type X-ray spectrum of black hole candidates is produced in the hot optically thin region of the disk, in which ions have temperature significantly exceeding electron temperature and give a dominant contribution to the pressure. Shapiro et al. /2/ pointed out that such "two-temperature" region could be formed in the inner part of the standard cool disk at high accretion rates, $\dot{m} \geq 1/110 \, (\alpha m_*)^{-1/8}$, at which the radiation pressure becomes equal to the gas pressure /1/. They supposed that the secular /29/ and thermal /30/ instabilities present in the radiation-pressure dominated part of the cool disk swell it to the hot gas-pressure dominated optically thin two-temperature region. The data of X-ray observations considered above can not be explained in the framework of this model. The observations unambiguously show that the disk becomes really hot (the hard state) only at small accretion rates and that this hot region should occupy the extended part of the disk (hard X-rays take away at least a half of the total disk luminosity). The "two-temperature" solution formally valid outside the inner disk region as well as within it can still be the best solution to explain the observed spectrum and the structure of the hard state. However the reason why the disk turns out to be in the hot state rather than in the cool one in this case is uncertain.

The hard spectral component observed from black hole candidates in the very high state differs drastically from that in the low state. It is steeper ($n \sim 2.5$), contains only a small ($\leq 10\%$) portion of the total disk luminosity and is likely formed within the disk region restricted in size and being under physically distinct conditions. We assume that in this case hard X-rays could be produced in the innermost region of the standard cool disk. According to Shakura & Sunyaev /1/ an optically thin region arises within the radiation-pressure dominated part of the disk at extremely high accretion rates, $\dot{m} \geq 1/14 \, m_*^{-1/32} \alpha^{-17/32}$. It remains compact ($r \leq 90 \, r_g \, \dot{m}^{64/93} m_*^{2/93} \alpha^{34/93}$, where r_g is the black hole gravitational radius) even when the accretion rate is close to the Eddington limit. The observed hard X-ray spectrum could be formed in this high temperature region due to Comptonization of low-energy photons. Finally, in our opinion accretion disk models could be reconciled with X-ray observations as follows:

- *very high/two-component spectral state* ($\dot{m} \geq 1/14 \, m_*^{-1/32} \alpha^{-17/32}$) – The disk consists of the cool optically thick extended outer region producing soft X-rays with the modified black-body spectrum (Thomson scattering dominates in opacity) and the hot optically thin radiation-pressure supported inner region producing hard X-rays.

- *high/soft state* – The whole disk is cool and optically thick. In its central part, $r \leq 580 \, r_g \, \dot{m}^{16/21} (\alpha m_*)^{2/21}$, the radiation pressure exceeds the gas one whenever $\dot{m} \geq 1/110 \, (\alpha m_*)^{-1/8}$ /1/. Near the outer boundary of this region the change in the disk surface temperature should take place reflecting the change in the disk structure. When $\dot{m}$ fall below the above value, e.g. in the case of X-ray novae, the radiation-pressure dominated region disappeares and this can result in the "kick" observed in the X-ray nova light curves.

- *low/hard state* – The whole disk (or its significant part) is hot and optically thin supported by the ion pressure and keeping the electron temperature nearly independent of r, $\dot{m}$ and α (the "two-temperature" solution).

- *off-state* – There is no available information now. The "off-state" could mean the source transition to another spectral state, e.g. again to the cool standard disk emitting in the unobservable soft X-ray and EUV bands (it seems that the results of Nova Cygni 1989 observations in quiescence /31/ support this point of view).

We are still far from understanding of all the details of disk accretion onto a black hole. Reporting here very briefly the results of the existing accretion disk models comparison with X-ray observations and giving in outline the possible scenario for disk accretion, we intend to encourage the future development of accretion disk models.

This research was supported in part by the Russian National Fundamental Research Foundation grant 94-02-05068a and ISF grant M8J000.

REFERENCES

1. N.I.Shakura & R.A.Sunyaev, *Astron. Astrophys.*, 24, 337 (1973).

2. S.L.Shapiro, A.P.Lightman & D.M.Eardley, *Astrophys. J.*, 204, 187 (1976).

3. R.A.Sunyaev & J.Trümper, *Nature*, 279, 506 (1979).

4. J.C.Ling, W.A.Mahoney, Wm.A.Wheaton & A.S.Jacobson, in: *20th Inter. Cos. Ray Conf.*, Moscow, Nauka, 1987, 1, 54.

5. H.Tananbaum, H.H.Gursky, E.Kellog, R.Giacconi, & C.Jones, *Astrophys. J.*, 177, L5 (1972).

6. S.Grebenev, et al., *Astron. Astrophys. Suppl. Ser.*, 97, 281 (1993).

7. R.Sunyaev, et al. *Astron. Astrophys.*, 247, L29 (1991).

8. M.N.Pavlinsky, S.A.Grebenev & R.A.Sunyaev, *Astrophys. J.*, 425, 110 (1994).

9. S.A.Grebenev, M.N.Pavlinsky & R.A.Sunyaev, *Adv. Space Res.*, this issue.

10. N.E.White & F.E.Marshall, *Astrophys. J.*, 281, 354 (1984).

11. S.Grebenev, R.Sunyaev, M.Pavlinsky, & I.Dekhanov, *Sov. Astr. Lett.*, 17, 985 (1991).

12. S.Miyamoto, et al., *Astrophys. J.*, 383, 784 (1991).

13. S.Grebenev, R.Sunyaev & M.Pavlinsky, in: *Workshop on Nova Muscae 1991*, ed. S.Brandt, DSRI, Lyngby, 1992, p.19.

14. J.E.McClintock & R.A.Remillard, *Astrophys. J.*, 308, 110 (1986).

15. J.Casares, P.A.Charles & T.Naylor, *Nature*, 355, 614 (1992).

16. R.A.Remillard, J.E.McClintock & C.D.Bailyn, *Astrophys. J.*, 399, L145 (1992).

17. K.Ebisawa, et al., *Astrophys. J.*, submitted (1994).

18. R.Sunyaev, et al., *Astrophys. J.*, 389, L75 (1992).

19. N.E.White, *Adv. Space Res.*, 3, 9 (1983).

20. Y.Tanaka, in: *23rd ESLAB Symp. on Two-Topics in X-Ray Astronomy*, ed. J.Hunt & B.Battrick, ESA SP-296, 1992, v.1, p.3.

21. V.Efremov, et al., in: *23rd ESLAB Symp. on Two-Topics in X-Ray Astronomy*, ed. J.Hunt & B.Battrick, ESA SP-296, 1992, v.1, p.15.

22. R.Sunyaev, et al., *Astron. Lett.*, 20, 890 (1994).

23. R.Sunyaev, et al., *Sov. Astron. Lett.*, 17, 123 (1991).

24. R.Sunyaev, et al., *Astron. Astrophys.*, 280, L1 (1993).

25. R.M.Hjellming & M.Rupen, *IAU Circ.*, 6060, 6073, 6086 (1994).

26. R.A.Sunyaev & L.G.Titarchuk, *Astron. Astrophys.*, 86, 121 (1980).

27. A.A.Galeev, R.Rosner & G.S.Vaiana, *Astrophys. J.*, 229, 318 (1979).

28. F.Haardt & L.Maraschi, *Astrophys. J.*, 413, 507 (1993).

29. A.P.Lightman & D.M.Eardley, *Astrophys. J.*, 187, L1 (1974).

30. N.I.Shakura & R.A.Sunyaev, *MNRAS*, 175, 613 (1976).

31. R.M.Wagner, et al., *Astrophys. J.*, 429, L25 (1994).

 Pergamon

Adv. Space Res. Vol. 19, No. 1, pp. (1)25–(1)28, 1997
© 1997 COSPAR
Printed in Great Britain. All rights reserved
0273–1177/97 $17.00 + 0.00

PII: S0273-1177(97)00032-X

LATEST COMPTEL RESULTS ON GALACTIC BLACK HOLE CANDIDATES

M. McConnell*, K. Bennett†, W. Collmar**, R. van Dijk***,‡,
D. Forrest*, W. Hermsen***, R. Much†, J. Ryan*,
V. Schönfelder**, H. Steinle** and A. Strong**

* *Space Science Center, University of New Hampshire, Durham NH, U.S.A.*
** *Max Planck Institut für Extraterrestrische Physik, Garching, Germany*
*** *SRON-Utrecht, Utrecht, The Netherlands*
† *ESTEC, Noordwijk, The Netherlands*
‡ *Astronomical Institute "Anton Pannekoek", University of Amsterdam, The Netherlands*

ABSTRACT

Since its launch in April of 1991, the *COMPTEL* experiment on the *Compton Gamma-Ray Observatory* has surveyed the entire sky in the energy range of 0.75 − 30.0 MeV. Here we survey the latest results obtained from the *COMPTEL* data with respect to galactic black hole candidates. The most prominent such object, Cygnus X-1, is also one of the brightest sources observed by *COMPTEL*. In addition, we report on the progress of a full survey of the sky, searching for evidence of MeV emission from other black hole candidates.
©1997 COSPAR. All rights reserved

INTRODUCTION

The *COMPTEL* instrument was designed to image celestial γ-rays in the energy range of 0.75 − 30.0 MeV. With a field-of-view of ∼ 1 steradian, *COMPTEL* is capable of imaging a large part of the γ-ray sky at any given time with a point source location accuracy of typically ∼ 1°. A more detailed description of the experiment can be found in Schönfelder et al. /1/.

CYGNUS X-1

The black hole candidate Cygnus X-1 is one of the strongest sources observed by *COMPTEL*. During the 3 years of the *CGRO* mission, it has been observed several times within the field-of-view of *COMPTEL*. Previously we have reported on the results of the analysis of data obtained during the first year of the *CGRO* mission /2/. These data indicated that: 1) there was no evidence for any hardening of the spectrum in the region near 1 MeV; and 2) the plasma temperature suggested by a Wien spectral model (the high energy limit of the Sunyaev-Titarchuk Comptonization model) was much higher than that generally implied by hard X-ray observations. This latter conclusion seems to require a revision in the standard spectral model for Cygnus X-1. In particular, it would suggest either a limitation in the standard Comptonization model or the need to incorporate a backscatter component resulting from the scattering of hard X-rays off a cooler (optically-thick) component of the accretion flow.

We have continued our analysis of Cygnus X-1, incorporating new observations as they become available. The latest analysis of Cygnus X-1 now incorporates all observations through February, 1994. The most recent of these observations (Viewing Period 318.1) was a target-of-opportunity observation of Cygnus X-1, prompted by an extremely low state of hard X-ray emission /3/. No positive detection was made by *COMPTEL* during this observation, despite a favorable exposure.

A summary of the observations used in the present analysis is given in Table 1, which also includes the average 0.75 − 2.0 MeV flux for each observation period. These data show that, although there are variations in the measured flux near 1 MeV (by more than a factor of two), these variations are not statistically significant. (We note that this is contrary to an earlier analysis, which had

Viewing Period	Start Date	End Date	Viewing Angle	0.75-2.0 MeV Flux $(cm^{-2}s^{-1}MeV^{-1})$
2.0	30-May-1991	8-Jun-1991	2°	$3.3(\pm1.0) \times 10^{-4}$
7.0	8-Aug-1991	15-Aug-1991	11°	$5.1(\pm1.4) \times 10^{-4}$
203	1-Dec-1992	22-Dec-1992	7°	$2.7(\pm0.7) \times 10^{-4}$
212.0	9-Mar-1993	23-Mar-1993	15°	$2.9(\pm0.9) \times 10^{-4}$
302.0	7-Sep-1993	9-Sep-1993	18°	
303.2	22-Sep-1993	1-Oct-1993	18°	
303.7	17-Oct-1993	19-Oct-1993	18°	
302-303				$2.3(\pm1.2) \times 10^{-4}$
318.1	1-Feb-1994	8-Feb-1994	5°	$2.9(\pm1.7) \times 10^{-4}$

Table 1: Summary of Cygnus X-1 observations used in the present analysis.

indicated that the variations were significant /2/.)

Figure 1 shows the average photon spectrum derived from all observations up through February, 1994. There is evidence for significant emission extending well above 2 MeV, with a data point in the 2-5 MeV range at a significance level of 4.4σ. (However, the observed flux near 1 MeV is still well below that reported by HEAO-3 and several other balloon observations /2/.) A fit to the *COMPTEL* data using a Wien spectrum gives an electron temperature value of $kT \sim 190$ keV. This is consistent with the temperature values which were derived from an earlier analysis of the *COMPTEL* data /4/. Continuing observations (e.g., three more weeks of data collected in June, 1994) should permit an ever more accurate determination of the time-averaged spectrum near 1 MeV.

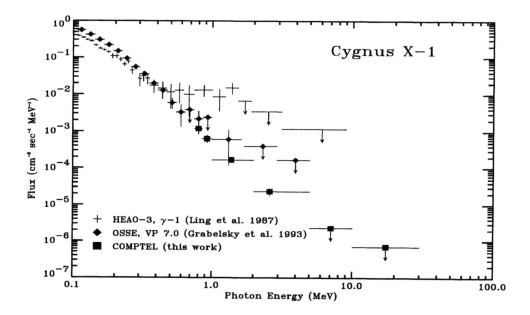

Figure 1: Average spectrum of Cygnus X-1 derived from all *COMPTEL* data collected through February, 1994 (Viewing Period 318.1). Also shown are representative spectra from both *HEAO-3* and *OSSE*.

OBSERVATIONS OF X-RAY TRANSIENTS

In order for sources to be observed by *COMPTEL*, they must lie within $\sim 30°$ of the *CGRO* pointing direction. Therefore, unless a given X-ray transient happens to lie within the *COMPTEL* field-of-view, a Target-of-Opportunity is usually required in order for COMPTEL to obtain useful data. To date, only two such Targets-of-Opportunity have been declared. The first was for Nova Persei (GRO J0422+32), which was discovered by *BATSE* in August of 1991. More than 3 weeks of data was obtained by *COMPTEL* within two months of the onset of the event. Preliminary results have been previously reported /4/. These data show evidence for emission in both the 0.75-1.0 MeV and 1-2 MeV energy intervals with some evidence for a hardening in the spectrum above 1 MeV. The second Target-of-Opportunity was that of GRO J1009-45. However, this observation lasted only about one day, so the *COMPTEL* exposure is quite poor. No emission was detected; the 2σ upper limits in the $0.75 - 1.0$ MeV and the $1 - 3$ MeV energy intervals are $1.5 \times 10^{-3} cm^{-2} s^{-1} MeV^{-1}$ and $3.1 \times 10^{-4} cm^{-2} s^{-1} MeV^{-1}$, respectively.

SYSTEMATIC SEARCH FOR OTHER BLACK HOLE CANDIDATES

The *COMPTEL* database (which covers the entire sky) affords the opportunity to search for emission from all known objects of a given class. We have therefore undertaken an effort to search for point sources which may be coincident with the known (or suspected) black hole candidates. For this purpose, we have generated a list of candidates based, in part, on that of Cowley [5].

Our initial work has concentrated on a search for time-averaged continuum emission using all of the data from phases 1 and 2 (the first 28 months) of the *CGRO* mission. Since this work is not yet completed, the results of this survey will be presented elsewhere. Here, in order to demonstrate the kind of results which are coming out of this study, we present two separate images of a wide field centered near Cygnus X-1. In both cases, we have plotted the locations of the known black-hole candidates within this field-of-view. (The plot symbols for each source are as follows: '+' = SS 433; '◇' = 1957+11; '△' = 2000+25; '□' = 2023+34.) The first image (Figure 2) shows the results we derive using a standard background analysis which is derived via a smoothing of the 3-dimensional COMPTEL dataspace /6/. The second image (Figure 3) incorporates a model which corresponds to the spatial distribution of molecular and atomic hydrogen (H_2 and HI, respectively) in the galaxy. The importance of modeling this diffuse galactic component at these energies is clear. Typical (2σ) upper limits for derived source fluxes in the $1 - 30$ MeV interval are on the order of a few times $10^{-6} cm^{-2} s^{-1} MeV^{-1}$.

Our future work will include the use of data from phase 3 and the search for phase-dependent emission from selected candidates. We have also begun a systematic search for 2.2 MeV line emission from X-ray binary systems. This is a bit more problematic due to a strong internally-generated 2.2 MeV background in *COMPTEL*. However, the origin of this background is well-known (neutron capture in the D1 detectors), and our recent efforts to model this background component appear quite promising.

ACKNOWLEDGEMENTS

This work was supported by NASA contract NAS5-26645, by the Bundesministerium für Forschung and Technologie under grant 50 QV 9096 and by The Netherland's Organization for Scientific Research (NWO).

REFERENCES

1. V. Schönfelder, H. Aarts, K. Bennett, H. de Boer, J. Clear, W. Collmar, A. Connors, A. Deerenberg, R. Diehl, A. v. Dordrecht, J.W. den Herder, W. Hermsen, M. Kippen, L. Kuiper, G. Lichti, J. Lockwood, J. Macri, M. McConnell, D. Morris, R. Much, J. Ryan, G. Simpson, M. Snelling, G. Stacy, H. Steinle, A. Strong, B.N. Swanenburg, B. Taylor, C. de Vries, and C. Winkler,

M. McConnell *et al.*

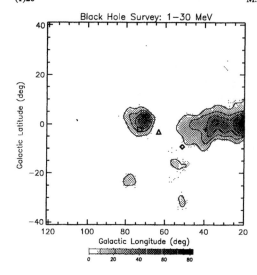
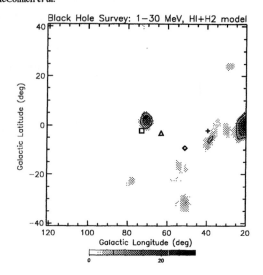

Figure 2: Image of the Cygnus region derived for 1-30 MeV energy range using a standard background analysis.

Figure 3: Same as in Figure 2, but with an additional background component which corresponds to the spatial distribution of diffuse emission in the galaxy.

Instrument Description and Performance of the Imaging Gamma-Ray Telescope *COMPTEL* aboard NASA's *Compton Gamma Ray Observatory, Ap. J. Supp.*, 86, 657 (1993).

2. M. McConnell, D. Forrest, J. Ryan, W. Collmar, V. Schönfelder, H. Steinle, A. Strong, R. van Dijk, W. Hermsen and K. Bennett, Observations of Cygnus X-1 by *COMPTEL* During 1991, *Ap. J.*, 424, 933 (1994).

3. B.A. Harmon, G.J. Fishman, R.C. Rubin, C.A. Wilson, and W.S. Paciesas, *IAU Circular No. 5881*, 20-Oct-1993.

4. R. van Dijk, H. Bloemen, W. Hermsen, W. Collmar, R. Diehl, J. Greiner, G.G. Lichti, V. Schönfelder, A. Strong, K. Bennett, L. Hanlon, C. Winkler, M. McConnell and J. Ryan, MeV Emission from the Black-Hole Candidate GRO J0422+32 Measured with *COMPTEL*, in: *The Second Compton Symposium*, eds. C.E. Fichtel, N. Gehrels, J.P. Norris, AIP, New York, p. 197.

5. A. Cowley, Evidence for Black Holes in Stellar Binary Systems, *Ann. Rev. Astron. Astrophys.*, 30, 287 (1992).

6. H. Bloemen, W. Hermsen, B.N. Swanenburg, C. de Vries, R. Diehl, V. Schönfelder, H. Steinle, A.W. Strong, A. Connors, M. McConnell, D. Morris, G. Stacy, K. Bennett, C. Winkler, *COMPTEL* Imaging of the Galactic Disk and the Separation of Diffuse Emission and Point Sources, *Ap. J. Suppl.*, 92, 419 (1994).

7. A.W. Strong, K. Bennett, H. Bloemen, R. Diehl, W. Hermsen, D. Morris, V. Schönfelder, J.G. Stacy, C. de Vries, M. Varendorff, C. Winkler, and G. Youseffi, Diffuse Continuum Gamma-Rays from the Galaxy Observed by COMPTEL, *Astron. Astr.*, in press (1994).

Pergamon

Adv. Space Res. Vol. 19, No. 1, pp. (1)29–(1)34, 1997
© 1997. Published by Elsevier Science Ltd on behalf of COSPAR
Printed in Great Britain. All rights reserved
0273–1177/97 $17.00 + 0.00

PII: S0273-1177(97)00033-1

THREE HARD X-RAY TRANSIENTS: GRO J0422+32, GRS 1716-24, GRS 1009-45. BROAD BAND OBSERVATIONS BY ROENTGEN-MIR-KVANT OBSERVATORY

A. S. Kaniovsky*, V. A. Arefiev*, N. L. Aleksandrovich*, V. V. Borkous*, K. N. Borozdin*, V. V. Efremov*, R. A. Sunyaev*, E. Kendziorra**, P. Kretschmar**, M. Kunz**, M. Maisack**, R. Staubert**, S. Doebereiner***, J. Englhauser***, W. Pietsch***, C. Reppin***, J. Truemper***, G. K. Skinner†, A. P. Willmore†, A. C. Brinkman‡, J. Heise‡ and R. Jager‡

Space Research Institute, Russian Academy of Sciences, Profsoyuznaya str. 84/32, Moscow, Russia
**Astronomisches Institut der Universitat Tubingen, Waldhauserstr. 64, W-7400 Tubingen, Germany*
****Max-Plank-Institut fur Extraterrestrische Physik, W-8046 Garching, Germany*
†*School of Physics and Space Research, University of Birmingham, Edgbaston, Birmingham, U.K.*
‡*Laboratory for Space Research, Utrecht, The Netherlands*

ABSTRACT

Three bright X-ray transient sources GRO J0422+32, GRS 1009-45 and GRS 1716-24 detected in 1992-1993 continue the sequence of such transient sources which were observed by the broad-band MIR-KVANT X-ray observatory in the period since 1988. Two of them, GRO J0422+32 and GRS 1716-29, may be classified as hard X-ray novae where the X-ray spectrum formation is probably dominated by comptonization, GRS 1009-45 is a typical soft X-ray nova with a strong soft thermal component and hard power-law tail. In this paper the results of broad-band (from 2 to 200-500 keV) spectroscopy for these sources are presented and are compared to earlier observations of other X-ray novae.

INTRODUCTION

The MIR-KVANT observatory is an international X-ray observatory which is able to perform X-ray observations in the 2-800 keV energy band. A detailed description of the observational capabilities can be found in /1/. We here discuss observations made in the period 1992-1993.

GRO J0422+32

This source was discovered by the BATSE experiment on board the Compton Gamma Ray Observatory (GRO). It is well studied in X-rays now due to the efforts of the GRO, GRANAT and ROSAT scientists. Optical observations /2/ proved the source to be a strong black hole candidate with a mass of 2.9 - 6.2 $M_\odot$. The evolution of the source fits the behavior of the other X-ray novae with the exception of some optical outbursts which were detected several times after the X-ray source flux fell below the limit of detection by X-ray observatories. ASCA observations of the source in the period of optical outburst in August,1993 showed that the X-ray source was very weak ~1 mCrab /3/.

The KVANT results of GRO J0422+32 observations were published in /4/ and can be summarized as follows:

1. The source spectra in the standard energy band 2-30 keV can be represented by a simple power law with photon index -1.5.

2. No soft absorption in the spectra was detected. The 3 σ upper limit for N$_{HI}$ according our data is 4 10^{21} cm^{-2}.

3. Broad band spectra in the range 2-150 keV can be satisfactory approximated by a single temperature comptonization model ST80 /5/ with the temperature 28-30 keV and τ = 1.9-2.

4. At the highest energies (150-500 keV) the data from HEXE and Pulsar X-1 show hard excess in comparison with the ST80 model. This agrees with the fact that the ST80 approximation for harder experiments, for example SIGMA, gives higher temperature and lower optical depth than our observations. The single temperature ST80 approximation can be characterized by the power law asymptotic at the low energies and exponential cutoff starting approximately at the energies of 3 kT$_e$. As the fact the softer experiments are more sensitive to the low energy power law and as a result more restricted in the possible temperature range because they detect the beginning of deviations from this asymptotic due to the thermal nature of the comptonized spectrum. Hard X-ray experiments fit the behavior of spectrum in the range of the exponential cutoff and are less sensitive to the power law part of the spectrum. Thus discrepancies in the results of experiments in different energy bands lead to the conclusion that a simple ST80 approximation is not adequate in the case of GRO J0422+32, and more detailed models for the spectrum formation such as the two-component comptonization model /6/ are necessary. The difference may also be due to deviations of the analytical approximation ST80 from reality due to relativistic effects and a deviation from the diffuse approach in the radiation transfer solution see /7/.

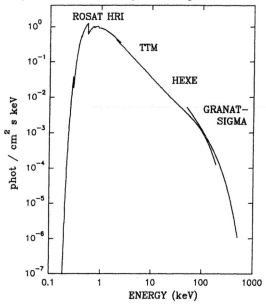

Fig.1. Combined spectrum of GRO J0422+32.

MIR-KVANT observations of GRO J0422+32 were performed simultaneously with ROSAT observations /8/. At the time of the publication of the ROSAT results, the data from the MIR-KVANT observatory were not available and so there was no information about the source intensity in the standard X-ray band. Consequently various assumptions about the intrinsic absorption were discussed.

If we reconstruct the broad-band spectrum of the source using ROSAT and SIGMA data Fig. 1 it becomes obvious that the question about the intrinsic absorption in the source can be resolved. In this figure the power law spectrum (α=-1.5) obtained from the TTM data is extrapolated to the ROSAT energy band, convolved with interstellar absorption in the direction to the source and normalized to a ROSAT intensity of 0.3 Crab. Of course the result may be strongly dependent on our knowledge of the Crab spectrum in ROSAT energy band which is not absolute, but the agreement of the ROSAT and the MIR-KVANT data seems remarkably good. An interstellar absorption in the direction of the source 1.7 10^{21} cm^{-2} seems to be sufficient to explain the behavior of the spectrum. From Fig. 1 one can see also the deviations of the one temperature ST80 fits at HEXE hard energies and SIGMA soft energies. The two experiments give respectively: τ =1.97, kTe=30.0keV and τ= 1.04 kTe= 48.1keV

The variation of the source intensity in our observations was in agreement with a previously determined decay time of 44 day /9/. But on 1992 Aug. 29 the intensity of the source was approximately 10% higher than the general decay curve. This possibly demonstrates short time variability of the source corresponding to very low frequency noise. The spectral variability of the source is not strong (see Table 1) but the spectrum was a little bit harder on September 16-17. This agrees with the assumption about the two component nature of the GRO J0422+32 spectrum see /6/.

TABLE 1 Analytical approximations for GRO J0422+32 spectra.

Date 1992	Energy	Fit type	χ^2 per d.o.f.	Fit parameters	
29.08-2.09	2-20 keV	power law	1.0	photon index	1.49±0.03
				absorption $N_H l$	<6.7 10^{21} cm^{-2} (3 sigma)
				iron line equiv. width < 0.29 keV (3 sigma)	
	2-500 keV	bremsstrahlung	6.1	Te	119±4 keV
		comptonized disk	1.5	Te	28.0±0.4 keV
				τ-half optical depth	2.00±0.02
16.09-17.09	2-20 keV	power law	1.7	photon index	1.5±0.03
				absorption $N_H l$	<6.6 10^{21} cm^{-2} (3 sigma)
				iron line equiv. width <0.21 keV (3 sigma)	
	2-500 keV	bremsstrahlung	10.0	Te	119±5 keV
		comptonized disk	1.8	Te	30.0±0.5 keV
				τ-half optical depth	1.97±0.03
29.08-17.09	2-20 keV	power law	1.8	photon index	1.49±0.02
				absorption $N_H l$	<5 10^{21} cm^{-2} (3 sigma)
				iron line equiv. width <0.175 keV (3 sigma)	

GRS 1716-24

This source was discovered Sept. 25-26, 1993 during the Galactic Center observations by the SIGMA experiment /10/ and simultaneously by the BATSE experiment /11/. Practically simultaneously, the Galactic Center region was observed by the MIR-KVANT observatory and the source was detected by the TTM telescope. In September - October, 1993 the source was observed several times by the TTM telescope and twice in conjunction with the HEXE experiment.

TABLE 2 Analytical approximations for GRS 1716-24 spectra.

Date 1993	Energy	Fit type	Xi^2 per d.o.f.	Fit parameters	Flux at 5 keV 10^{-10} erg/s/cm2/keV
26.09	2-20 keV	power law	1.0	photon index 1.36±0.27	2.81±0.40
27.09	2-20 keV	power law	1.3	photon index 1.64±0.13	7.24±0.47
3.10	2-20 keV	power law	0.6	photon index 1.47±0.25	8.66±1.21
3.10	2-20 keV	power law	1.2	photon index 1.51±0.23	9.27±1.14
7.10	2-200 keV	bremsstrahlung	1.6	Te 111±6 keV	8.28±0.26
		comptonized disk	1.4	Te 22.4±0.7 keV	8.28±0.26
				α 0.563±0.016	
3.10	2-20 keV	power law	1.2	photon index 1.88±0.25	12.22±1.46
24.10	2-200 keV	bremsstrahlung	3.0	Te 94.5±4.4 keV	7.90±0.32
		comptonized disk	1.8	Te 25.8±0.7 keV	7.90±0.32
				α 0.668±0.015	

The broad band spectrum of the source is very similar to that of GRO J0422+32. No soft component was detected. The spectrum can be approximated with a simple comptonization model. Observations of the source with TTM can only give the power law slope for the soft part of the spectrum. No significant soft absorption was detected, but the upper limit according to our observations is rather high: ~1.5 10^{22} cm^{-2}

(3σ). All the results are presented in Table 2. According to the TTM data it is difficult to reach any conclusion about the variability of the spectral shape in the standard X-ray band. Nevertheless the broad band observations indicate some changes in the spectral shape, see Table 2 and Fig.2.

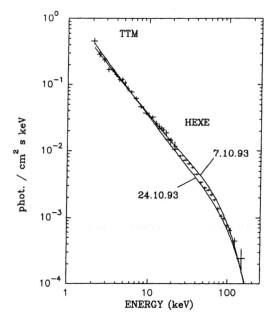

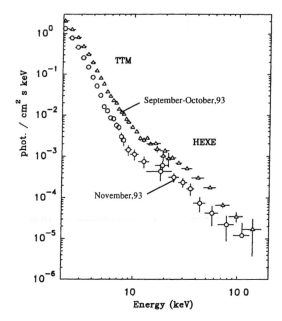

Fig.2. Mean broad-band spectrum of GRS 1716-24 according to MIR-KVANT observations. Comptonized disc approximations for different observations are plotted together with the mean photon spectrum (crosses).

Fig.3 Spectra of GRS 1009-45 according to MIR-KVANT observations in September-October and November, 1993.

GRS 1009-45

This source was discovered by the WATCH(GRANAT) instrument on 1993 Sept.11. The source was first detected in the hard energy band of the WATCH detector (20-60 keV) and only a day later in the soft energy band 8-20 keV /12/. After the first observations by MIR-KVANT it was shown that the source could be classified as a soft X-ray nova because of a strong soft thermal component in the spectrum together with a hard power law tail /13/. Simultaneous observations of the source with the GRO observatory instruments BATSE /14/ and OSSE /15/ in hard X rays show pure power law behavior of the source spectra up to energies of ~500 keV.

Continuous observations show that the spectrum can not be modeled solely by the superposition of a black-body and a power law spectra. At energies ~10 keV a lack of photons is detected see Fig. 3. This may be interpreted as strong iron absorption or by the involvement of another hard component in addition to the power law tail, such as reflection or a comptonized bremsstrahlung component.

In Fig. 3, the spectra of the source are presented for two long sets of observations in September-October, 1993 and November, 1993. It is clear that the temperature of the soft component decreases from 0.50± 0.01 to 0.45±0.01 keV while the power law slope remains unchanged α=-2.23± 0.03. The relative luminosity of the hard component decreases 1.6 times more then the soft one.

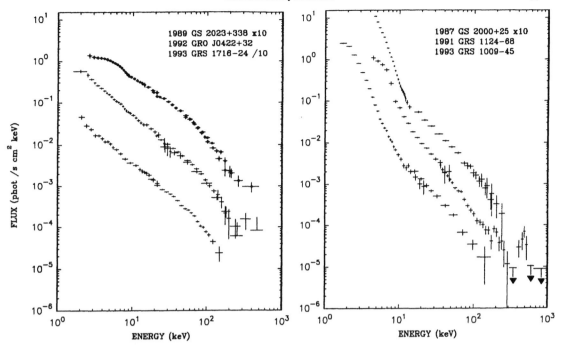

Fig.4. Spectra of hard X-ray transients observed by MIR-KVANT and GRANAT in 1988-1993.

DISCUSSION

Since 1988 MIR-KVANT observed 5 X-ray transients: GS 2000+25, GS 2023+338, GRO J0422+32, GRS 1716-24 and GRS 1009-45. Two other bright transients GRS 1124-684 and GRS 1915+105 were not observed because of difficulties with the observatory pointing. In Fig. 4 broad band spectra of the hard X ray transients are presented separately for hard X-ray novae and soft X-ray novae. The spectrum of GRS 1124-684 is plotted according GRANAT observations /16/. For GRS 1915+105 only the hard power law spectrum was measured by SIGMA for the outburst in August-September, 1992 (not plotted) /17/. In the observations by the TTM-MIR-KVANT during the recent outburst in 1994 Sept. a strong soft component was detected for GRS 1915+105 /18/.

There is a remarkable correlation such that those transients with a soft component in the spectrum, exhibit power law hard tails (see Fig.4) while at the same time hard X-ray transients without soft components have strong exponential cutoffs in the spectra at energies of the order 100 keV, typical of the comptonization powered hard spectra.

Concerning the spectral evolution of hard X-ray transients, it is necessary to mention that up to now we have no evidence of the softening of hard X-ray spectra (except GRS1915+105, where the radiation is probably of a different nature) On the contrary some sources demonstrate obvious hardening of the spectrum (GS2023+338, GRS1124-684, GRO J0422+32). This may be considered as a strong argument in favour of models involving Compton reflection mechanisms /19/ or in the case off the models involving multi-temperature electron distributions this probably points to a general correlation between the decay time and electron temperature.

Until recently the class of hard X-ray transients seemed more or less uniform from the point of view of light curve evolution and hard tails in the spectra. However, recent observations show that this is probably not the case. In particular the decay time of GRS 1716-24~300 days was considerably longer than that typical for previously detected transients and the source demonstrated a "turn off" behavior in the light curve, in which the intensity falls by an order of the magnitude in several days /20/. Detection of the strong radio emitting jets from GRS 1915+105 and from the newly discovered GRO J1655-40 /21/, points to the fact that here formation of hard X-ray spectra may be powered by synchrotron mechanism. These sources are also remarkable for demonstrating a series of outbursts and the shorter X-ray light curve decay times in comparison with other hard X-ray transients. So probably the general class of hard X-ray transients needs to be considered as a very broad one from the point of view of the physical nature of the X-ray sources.

REFERENCES

1. R.Sunyaev, A.Kaniovsky, V.Efremov, M.Gilfanov, E.Churazov, S.Grebenev, A.Kuznetsov, A.Melioransky, N.Yamburenko, S.Unin, D.Stepanov, I.Chulkov, N.Pappe, M.Boyarsky, E.Gavrilova, V.Loznikov, A.Prudkogliad, V.Rodin, C.Reppin, W.Pietsch, J.Englhauzer, J.Truemper, W.Voges, E.Kendziorra, M.Bezler, R.Staubert, A.C.Brinkman, J.Heise, W.A.Mels, R.Jager, G.K.Skinner, O.Al-Emam, T.G.Patterson, and A.P.Willmore, *Nature* , 330, N6145, 227, (1987).

2. T.Kato, S.Mineshige and R.Hirata, *IAU Circ.* 5704 (1992).

3. Y.Tanaka, *IAU Circ.* 5851, (1993).

4. R.Sunyaev, A.S.Kaniovsky, K.N.Borozdin, V.V.Efremov, V.A.Aref'ev, A.S.Melioransky, G.K.Skinner, H.C.Pan, E.Kendziorra, M.Maisack, S.Doebereiner, W.Pietsch, *Astron. Astrophys*, 280, L1, (1993).

5. R.Sunyaev and L. Titarchuk, *Astron. Astrophys.*, 86, 121, (1980).

6. A.Finoguenov et al., This issue, (1994).

7. L.Titarchuk, 1994 This issue, (1994).

8. W.Pietsch, F.Haberl, N.Gehrels, and R.Petre, 1993 *Astron. Astrophys.*, 273,L11, (1993).

9. A.Finoguenov, *IAU Circ.* 5608, (1992).

10. J. Balet, M.Denis, M.Gilfanov and R.Sunyaev, *IAU Circ.*5874, (1993).

11. B.A.Harmon, *IAU Circ.* 5874, (1993).

12. I.Lapshov, S.Sazonov, R.Sunyaev, S.Brandt, A.Castro-Tirado, N.Lund, *Sov. Astron. Letters*, 20, 250, (1994).

13. A.Kaniovsky, K.Borozdin and R.Sunyaev, *IAU Circ.* 5878 (1993).

14. B.A.Harmon, S.N.Zhang, C.A.Wilson, B.C.Rubin, G.J.Fishman, and W.S.Paciesas, BATSE observations of transient hard X-ray sources, *Proceedings of Compton Symposium September 20-22,* in press (1993).

15. J.Kurfess et al., This issue, (1994).

16. M.Gilfanov, E.Churazov, R.Sunyaev, S.Grebenev, M.Pavlinsky, A.Dyachkov, V.Kovtunenko, R.Kremnev, A.Goldwurm, J.Ballet, P.Laurent, J.Paul, E.Jourdian, M.C.Schimtz-Fraysse, J.P.Roques, and P.Mandrou , *Astron.Astrophys. Suppl. Ser.*, 97, 303, (1993).

17. A.Finoguenov, E.Churazov, M. Gilfanov, R.Sunyaev, A.Vikhlinin, A.Dyachkov, N.Khavenson, I.Tserenin, M.C.Schimtz-Fraysse, M.Denis, J.P.Roques, P.Mandrou, AClaret, J.Ballet, B. Cordier, and A.Goldwurm, *Ap.J.*,424,940, (1994a).

18. N.Aleksandrovich, K Borozdin and R.Sunyaev, *IAU Circ.* 6080, (1994).

19. S.Doebereiner, M.Maisack, J.Englhauser, W.Pietch, C.Reppin, J.Truemper, E.kendziorra, P.Kretschmar, M.Kunz, R.Staubert, V.Efremov, A.Kaniovsky, A.Kuznetzov, and R.Sunyaev, *Astron.Astrophys.*,287,105, (1994).

20. B.A.Harmon, *IAU Circ.* 5913, (1993).

21. I.F.Mirabel and L.F.Rodriges, *Nature* 371,46, (1994).

 Pergamon

Adv. Space Res. Vol. 19, No. 1, pp. (1)35–(1)39, 1997
© 1997. Published by Elsevier Science Ltd on behalf of COSPAR
Printed in Great Britain. All rights reserved
0273–1177/97 $17.00 + 0.00

PII: S0273-1177(97)00034-3

GRANAT/SIGMA OBSERVATIONS OF X-RAY NOVA PERSEI 1992

A. Finoguenov*, M. Gilfanov*, E. Churazov*, R. Sunyaev*,
A. Vikhlinin*, A. Dyachkov*, V. Kovtunenko*, R. Kremnev*,
M. Denis**, J. Ballet**, A. Goldwurm**, P. Laurent**,
E. Jourdain***, J. P. Roques***, L. Bouchet*** and
P. Mandrou***

Space Research Institute, Russian Academy of Sciences, Profsouznaya 84/32, 117810 Moscow, Russia
**Service d'Astrophysique, Centre d'Etudes Nucleaires de Saclay, 91191 Gif-sur-Yvette Cedex, France*
***Centre d'Etude Spatiale des Rayonnements, 9 Avenue du Colonel Roche, BP 4346, 31029 Toulouse Cedex, France*

ABSTRACT

We report on the results of the *GRANAT/SIGMA* observations of the X-ray transient GRO J0422+32 in the 35–1300 keV energy band in 1992–93. In the maximum of the light curve, this black hole candidate was the brightest hard X-ray source detected by *SIGMA* so far. During the monitoring of the GRO J0422+32 outburst, *SIGMA* detected a gradual hardening of the source spectrum. In terms of thermal emission of optically thin plasma this hardening corresponds to an increase of the best-fit temperature from ~ 110 keV to ~ 140 keV. Source spectrum is well represented by two component Comptonized disk model /9/. The parameters of the model do not exhibit any statistically significant variation from the average value of $(\alpha;\ kT_e)$ of $(0.737 \pm 0.01;\ 25.4 \pm 0.05)$ for the softer component and $(0.39 \pm 0.09;\ 62.6 \pm 2.4)$ for the harder (with $\tau_{disk} \sim 1.8$ for both components). Nevertheless relative normalizations of the two components are changing during our observations by a factor of two. The harder component is fluctuating on a day time-scale and a growth of its contribution to the total flux is accompanied by an increase in the fractional RMS [0.001–0.1 Hz].

INTRODUCTION

From the initial outburst on Aug. 5 1992 (TJD=8839, we will further use $TJD = JD - 2,440,000.5$), detected by *BATSE* /6/ Persei Nova became one of the brightest objects ever seen at hard X-rays. The follow up *SIGMA* observations showed, that from 1992 August 15 till September 25 the 40-150 keV flux of the source decreased monotonically from 2.9 to 1.0 Crab following an exponential decay law with an e-folding time of 43 days /12/. In December 1992 GRO J0422+32 showed a strong secondary maximum, started at TJD=8957 /4/. In a subsequent *SIGMA* observation on February 7 1993 (TJD=9022) GRO J0422+32 was detected in the 40-150 keV energy band on the level of 135 ± 14 mCrab, i.e. ~ 3 times higher than is predicted from the 43-day exponential decay, that can be explained by the secondary maximum, observed by *BATSE*.

The overall flux history can be described by the model of /1/ in a pure mass instability form (we will designate it as "AKS model"). For *BATSE* observations this model gives T_0=8715 TJD, T=123 days, γ_m^{-1}=40 days, β^{-1}=5.5 days and $\alpha\psi$=0.09, respectively /5/ (see /1/ for parameters meaning).

According to *BATSE* data, the source spectrum experiences a softening at the time of initial rise (first ten days) and then a gradual hardening /7/.

Here we present an analysis of the *SIGMA* data, with emphasis to the source spectral behavior during 1992 Aug. 15 till Sep. 25 (after the first rise) and Feb. 7 1993 (after secondary maximum). In the 35-1300 keV band of the telescope operation the source was detected up to $\sim$600 keV. For a detailed telescope description, see /8/.

SIGMA RECORDS ON GRO J0422+32 FLUX HISTORY

As a good approximation of the GRO J0422+32 flux history, we used AKS model (we will refer here only to pure mass instability form) in the 40-600 keV energy band, with fixed T_0 and T from *BATSE* data, but allowing γ_m^{-1}, β^{-1} and $\alpha\psi$ to vary. The best-fit curve is shown in Fig.1 along with our data.

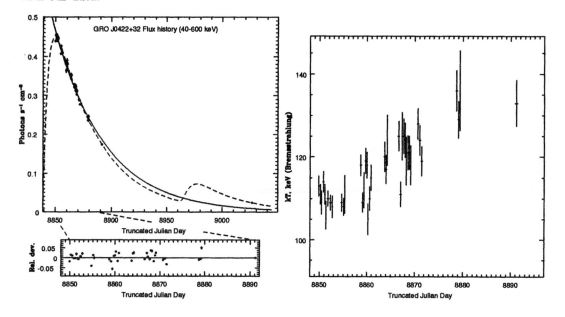

Figure 1. GRO J0422+32 flux history (40–600 keV) during 8849–9022 TJD (Aug.15, 1992 – Feb.7, 1993). Solid line: exponential decay law, dashed line: AKS model (parameter values are given in the text). Residuals are for AKS model.

Figure 2. kT$_e$, as derived from the thermal bremsstrahlung model, vs time.

From our data, γ_m^{-1}= 39.2 ± 0.6 days, β^{-1}= 4.6 ± 0.16 days and $\alpha\psi$= 0.13 ± 0.02 with χ^2 value much better then for a simple exponential decay law (best fit τ = 45 days) – 162/36 vs 335/37. Residuals from the best-fit curve of AKS model, as presented in Fig.1, show that source flux variations on a day time-scale although statistically significant, are rather small with maximal relative deviation less than 5% during our observations when the source flux drops by a factor of three. For the purposes of our analysis we will further use all our data except for the observation on Feb. 7 1993 (after secondary maximum) and fix $\alpha\psi$=0.09, according to *BATSE* data.

SPECTRUM

GRO J0422+32 spectrum in the 40–600 keV band possesses a clear cutoff, suggesting a choice of a thermal model for the source spectrum description. As a first approach, we used thermal bremsstrahlung. This model, although not physically meaningful, could provide a good representation of the overall source spectral shape and can be used for quantitative description of the source spectral evolution.

As could be seen from Fig.2, the best-fit bremsstrahlung temperature is constant during the first few days of our observation, with typical value of 110 keV and then gradually increases up to 130-140 keV. It generally follows the temperature behavior found by *BATSE*, although the temperature range is higher in *SIGMA* data, reflecting the difference in the energy ranges selected for the fit (20–300 keV for *BATSE* vs 40–600 keV for *SIGMA*).

We then build a more detailed analysis of the source spectrum, basing on the Comptonized disk model /9/ (ST80). As can be seen from Fig.3, the source spectrum in the 200-600 keV is clearly in excess of the extrapolation of the model, applied to the 40–150 keV energy range. The broad-band spectrum could not be satisfactory described in terms of this model with single temperature and an introduction of a second temperature component provides a better fit with confidence level more than 99.99% (from F-test) for each of our observations. From the physical point of view, our model approximates the temperature distribution in the Comptonization region by subdividing the emergent Comptonized spectrum into a softer (lower temperature) and a harder (higher temperature) components.

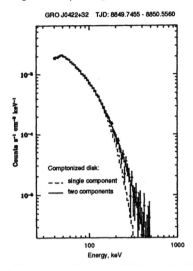

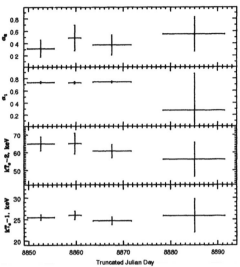

Figure 3. GRO J0422+32 count spectrum in the 40–600 keV energy band, measured by *SIGMA* on Aug.15. Dashed line represents the best-fit curve of ST80 model with single temperature derived in the 40–150 keV energy range, solid line corresponds to a ST80 model with two components applied to the spectrum in the 40–600 keV band.

Figure 4. $(\alpha; kT_e)$ parameters of the two component Comptonized disk model for GRO J0422+32 spectrum in the 40–600 keV energy band.

In Fig.4 we show $(\alpha; kT_e)$ parameters for both components, where we grouped observations to obtain a better statistics. The interesting result consists in the absence of any statistically significant variation of the parameters, while the normalizations are changing by a factor of three and a relative normalization varies by a factor of two, reflecting the previously mentioned hardening of the spectrum. The averaged parameters $(\alpha; kT_e)$ are $(0.737 \pm 0.01; 25.4 \pm 0.05$ keV$)$ for the softer component and $(0.39 \pm 0.09; 62.6 \pm 2.4$ keV$)$ for the harder. Values of τ_{disk}, derived from the best-fit parameters, are 1.78 ± 0.02 and 1.6 ± 0.3, for softer and harder temperature components respectively.

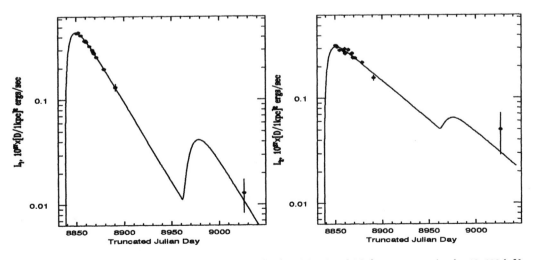

Figure 5. GRO J0422+32 light curves for softer *(left)* and harder *(right)* components in the 40–600 keV energy range.

A. Finoguenov *et al.*

DETAILS OF THE SOURCE SPECTRAL EVOLUTION

We then fixed the α and kT_e parameters and derived the light curves for the softer and harder component (shown in Fig.5). In using AKS model, the $(\gamma_m^{-1}; \beta^{-1})$ are $(28.5 \pm 1.6$ days; 5.7 ± 0.5 days) for the softer component and $(58 \pm 6$ days; 4.8 ± 0.8 days) for the harder, with the difference mainly in γ_m^{-1}.

It is interesting that the softer component does not show any significant variation from the best-fit AKS model ($\chi^2 = 7.8/11$) while the harder does ($\chi^2 = 20.0/11$).

Using the best fit parameters of the model it is possible to estimate the total energy, emitted in either components (in the 40–600 keV band). Expressed in terms of the mass, accreted on the compact object these correspond to $1.60 \pm 0.2 \times 10^{-10} M_\odot$ and $1.92 \pm 0.2 \times 10^{-10} M_\odot$ for the softer and harder components (assuming accretion efficiency of 0.057 and distance to GRO J0422+32 of 1 kpc). The amount of mass involved in the secondary maximum is then 9% of m_0 for GRO J0422+32.

Timing analysis of the *SIGMA* data in the 40–150 keV energy range showed that the source has large fractional RMS /11/. We then compare the behavior of the harder component with evolution of the fractional RMS [0.001–0.1 Hz] with time.

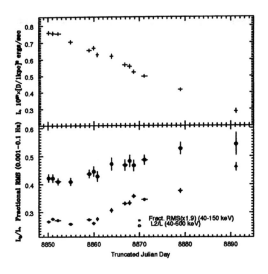
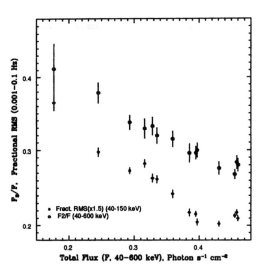

Figure 6. GRO J0422+32 luminosity evolution (40–600 keV), relative impact of the harder component to total luminosity and the fractional RMS [0.001–0.1 Hz] (multiplied by 1.9) vs time (TJD)

Figure 7. Relative contribution of the harder component to the GRO J0422+32 flux in the 40–600 keV band and the fractional RMS [0.001–0.1 Hz] (multiplied by 1.5) vs total source flux in the 40–600 keV energy range.

In Fig.6 it is shown that with decreasing overall luminosity (40–600 keV), the fractional RMS [0.001–0.1 Hz] grows, following changes in relative luminosity of the harder component. In Fig.7 we express a dependence of fractional RMS and a contribution of the harder component in terms of the source flux (40–600 keV). The growth of fractional RMS with hardening of the spectrum was also recently found in Cyg X-1 /2/.

CONCLUSIONS

The flux history of the GRO J0422+32 is well represented by AKS model with maximal relative deviation less than 5%. Source spectrum hardens with decreasing luminosity and the source spectrum in the 40–600 keV energy band could not be satisfactory described in terms of single component Comptonized Disk model (ST80). With the introduction of a second temperature component, model provides a better fit to data. The parameters of this model are constant with $(\alpha; kT_e)$ equal to $(0.737 \pm 0.01; 25.4 \pm 0.05$ keV) for the softer component and $(0.39 \pm 0.09; 62.6 \pm 2.4$ keV) for the harder. The luminosity evolution in the 40–600 keV band of each component is characterized by different exponential decay times (~ 30 days vs ~ 60 days), but the integrated luminosities are the same, requiring in sum $\sim 3.5 \times 10^{-10} M_\odot$ to be involved in the initial outburst (in assumption of 1 kpc source distance and 5.7% accretion efficiency). The harder component exhibits larger fluctuations than the softer on a time scale of days and the fractional RMS [0.001–0.1 Hz] increases with growth of the relative contribution of the harder component.

ACKNOWLEDGMENTS

The research described in this paper was made possible in part by Grant M7M000 from the International Science Foundation. A.Finoguenov acknowledges support from the International Science Foundation in participation at the conference.

REFERENCES

1. T.Augusteijn, E.Kuulkers and J.Shaham, A&A, preprint, (1993). (AKS)

2. M.Gilfanov et al., these proceedings.

3. M.Grebenev et al., A&A S, 97, 281, (1993).

4. B.A.Harmon et al., IAUC 5685, (1992)

5. B.A.Harmon et al., Compton Observatory Symposium, Maryland, (1993).

6. W.S.Paciesas et al., IAUC 5580, (1992)

7. W.S.Paciesas et al., The Integral Workshop: the multi-Wavelength Approach to Gamma-Ray Astronomy, Les Diablerets, Switzerland, (1993).

8. J. Paul et al. Advances In Space Research, 11, 289, (1991).

9. R.Sunyaev and L.Titarchuk, A&A, 86, 121, (1980). (ST80)

10. R.Sunyaev et al., A&A, 280, L1 (1993).

11. A.Vikhlinin et al., ApJ, to appear in March 10 issue, (1995).

12. A.Vikhlinin et al., IAUC 5608, (1992).

Adv. Space Res. Vol. 19, No. 1, pp. (1)41–(1)44, 1997
© 1997 COSPAR
Printed in Great Britain. All rights reserved
0273–1177/97 $17.00 + 0.00

PII: S0273-1177(97)00035-5

THE DISCOVERY OF 8.7 S PULSATIONS FROM THE ULTRASOFT X-RAY SOURCE 4U0142+61

G. L. Israel*,†, S. Mereghetti** and L. Stella***

*International School for Advanced Studies (SISSA-ISAS), Via Beirut 2-4, I-34014 Trieste, Italy
** Istituto di Fisica Cosmica del CNR, Via Bassini 15, I-20133 Milano, Italy
*** Osservatorio Astronomico di Brera, Via Brera 28, I-20121 Milano, Italy
† Affiliated to I.C.R.A.

ABSTRACT

We discovered a periodicity at $\sim$8.7 s from the X–ray sources 4U0142+61, previously considered a possible black hole candidate on the basis of its ultrasoft spectrum. The pulsations are visible only in the 1–4 keV energy range, during an observation obtained with the EXOSAT satellite in August 1984. In the same data, periodic oscillations at 25 minutes had been previously found in an additional hard spectral component dominating above 4 keV. A search for delays in the pulse arrival times caused by orbital motion gave negative results. Though the very high ($>10^4$) X–ray to optical flux ratio of 4U0142+61 is compatible with models based on an isolated neutron star, the simplest explanation involves a low mass X–ray binary with a very faint companion, similar to 4U1626–67. The discovery of periodic pulsations from 4U0142+61 weakens the phenomenological criterion that an ultrasoft spectral component is a signature of accreting black holes.
©1997 COSPAR. All rights reserved

INTRODUCTION

The persistent X–ray source 4U0142+61 was discovered by Uhuru and soon noticed to possess an ultrasoft spectrum. In the X–ray colour–colour diagram (/1/) it occupies the same region of black hole candidates in their "high state", such as LMCX–3, LMCX–1 and A0620–00. 4U0142+61 lies in the galactic plane (l=129".4, b=–0".4) and, despite its small error circle (a few arcseconds), no optical or radio counterparts have yet been identified (/2/).

During one of three EXOSAT observations of 4U0142+61 carried out in 1984–1985, an additional spectral component was detected above 3 keV within the $\sim$90' collimator response of the Medium Energy (ME) experiment. Correspondingly, $\sim$25 min periodic oscillations were discovered in the 3–10 keV energy range (/2/). This spectral component and periodicity were suggested to arise from RX J0146.9+6121, an X–ray transient discovered with ROSAT and identified with the Be star LSI +61" 235 (/3/; /4/).

Here we present the results of a re-analysis of the EXOSAT data, which led to the discovery of 8.7 s coherent pulsations in the 1–4 keV band (/5/ and /6/).

TIMING ANALYSIS

1984 EXOSAT ME Observation

During the $\sim$12 hr observation of August 27–28, 1984 the average 1–10 keV count rate of 4U0142+61 in the EXOSAT ME experiment was $\sim$10.7 counts s^{-1}. Due to the presence of the 3-10 keV spectral component showing the 25 min oscillations, this rate was ∼40% higher than that measured during the 1985 November 11 and December 11 observations, when the additional component was not present. In the 1984 observation, the ME instrument provided arrival time corrected light curves with a time resolution of 1 s in different energy bands. The power spectrum of the 1–3 keV light curve over an interval of 32768 s (Fig. 1) revelead the presence of two highly significant peaks (random occurrence probabilities of 8.9×10^{-8} and 3.6×10^{-9}, respectively), at the fundamental and the second harmonic of a coherent modulation with a period of 8.6872 s. The power spectrum of the 4–11 keV light curve did not show any evidence for these pulsations. On the contrary, the peaks corresponding to the first three harmonics of the 25 min modulation were clearly visible in the 4–11 keV power spectrum, but not in the 1–3 keV one (Fig. 1).

A more precise pulse period measurement of 8.68723 ± 0.00004 s and an upper limit to the period derivative of $\dot{P} < 6.2 \times 10^{-9}$ were obtained by using a phase fitting technique.

Between 1 and 4 keV the folded light curve consists of two peaks separated in phase by ∼$0.4 - 0.5$, with a peak to peak amplitude of ∼15%.

A search for an orbital modulation of the arrival times of the 8.7 s pulses was carried out for orbital periods ranging from 430 s to 43000 s No significant modulation was found. The 99% confidence upper limit to $a_x \sin i$ was derived to be 0.37 lt s (for a circular orbit).

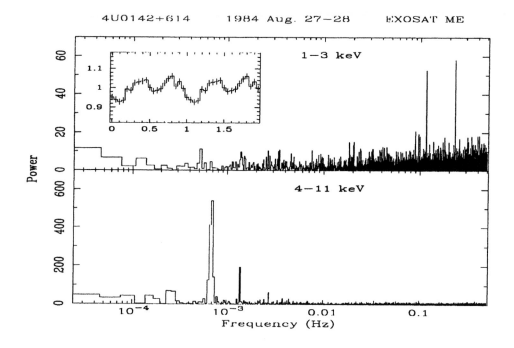

Fig. 1. 1984 EXOSAT ME observation power spectra in two different energy ranges. The peaks corresponding to the first two harmonics of the ∼8.7 s pulsations are clearly visible in the 1–3 keV power spectrum (upper panel). The three peaks in the 4–11 keV power spectrum testify to the presence of the ∼25 min modulation in that energy range.

Other Observations

During the two 1985 EXOSAT observations (November 11 and December 11), energy–resolved ME data were obtained only with an integration time of 10 s and did not provide useful information regarding the presence of the 8.7 s pulsations. Only the summed rates from the ME Argon and Xenon chambers were available with a time resolution of 0.25 s. Based on these data, we accumulated 1 s resolved light curves, which, however were characterised by a very high level of counting statistics noise due to the high background from the Xenon chambers (∼250 ct s⁻¹). The periodicity was searched in these light curves using a folding technique. Only a marginal evidence for the coherent pulsations at ∼8.7 s was obtained from the 1985 observations (see ref. /6/).

During the three EXOSAT ME observations, simultaneous data in the 0.05-2 keV band were obtained with the low energy (LE) telescope. The folded LE light curve of 4U0142+61 provided no evidence for the 8.7 s modulation (upper limit of 25%, 34% and 44% with 99% confidence).

A 2180 s ROSAT observation of 4U0142+61, was carried out on February 13, 1991 with the HRI instrument (0.1–2.4 keV). The arrival times of the 2846 counts within a radius of 20 arsec from the position of 4U0142+61 were searched for periodicities. Again we obtained only marginal evidence for pulsations at ~8.7 s.

Even in the absence of a clear detection with an X–ray imaging instrument, we concluded (/6/) that the ~ 8.7 s pulsations are most likely produced by the ultrasoft source 4U 0142+61.

This interpretation has been recently confirmed by using a 10700 s ROSAT PSPC observation carried out on 1993 February 12–13, which revealed the presence of ~ 8.7 s pulsations in the X–ray flux of 4U0142+61 (/7/). The heliocentric period was 8.6878 ± 0.0001 s with peak to peak amplitude of ~10%, close to the values worked out from the 1984 EXOSAT ME data and implying a spin down timescale of ~ 10^5 yr. We note that a period so close that of the 1984 observation argues against the validity of the marginal detections in the 1985 and 1991, which, on the contrary, implied a spin-up time of ~ 540 yr relative to the 1984 period measurement.

During the same observation imaging data from the transient X–ray source RX J0146.9+6121 were also obtained, which showed a strong ~60% modulation with a period of 1413 ± 8 s, confirming that this source is responsible for the 25 min modulation revealed in the 1984 EXOSAT ME and that it contains the slowest rotating neutron star known (in view of the spin up timescale of ~ 300 yr which rules out a white dwarf).

DISCUSSION

The lack of an optical counterpart down to limits of V~24 and R~22.5 (/2/) implies an X–ray to optical flux ratio, $F_x/F_{opt} > 10^1$. The only known classes of galactic sources which can yield such a high F_x/F_{opt} value are low mass X–ray binaries (LMXRBs) and isolated neutron stars.

A Low Mass X–ray Binary ?

A magnetic neutron star accreting from a low mass companion is the most likely model for 4U0142+61. Coherent pulsations are rarely seen in LMXRBs: the only known examples among optically identified systems, 4U1626–67, HerX–1 and GX1+4, have very different X–ray properties, companion stars and evolutionary origins. The spin period of 4U0142+61 is very similar to that of 4U1626–67 (7.7 s), and it is interesting to note that two other optically unidentified pulsars, which are likely accreting from low mass companions, 1E2259+586 and 1E1048.1–5937, have periods of the same order: 6.98 and 6.44 s respectively (/8/; /9/).

While the position of 4U0142+61 is close (<0.5") to that of two open clusters with well determined distance and reddening (/10/; NGC654, 2.5 kpc and NGC663, 2.1 kpc) , the column density N_H ~1.5×10^{22} cm^{-2} derived from the power law spectral fits of 4U0142+61 (/2/) corresponds to a greater distance. However, 4U0142+61 is not necessarily much further than these clusters, since a part of its absorption could be intrinsic or due to a local (d<1 kpc) molecular cloud which is present in this region. 4U0142+61 lies near the edge of this cloud, which does not significantly affect NGC654 and NGC663 (/10/). A 4 kpc distance would yield to a 1–10 keV luminosity of ~10^{36} ergs s^{-1}, similar to 4U1626–67. At this distance and reddening the faint optical counterpart of the latter source would be fainter than the present limits for 4U0142+61. On the other hand, an evolved companion similar to that of GX1+4 or CygX–2 would have been detected even at ~10 kpc. A companion star similar to that of 4U1626–67, i.e. either a main sequence star with M≃$0.08M_\odot$ or a white dwarf of 0.02 $M_\odot$ (/11/), is also compatible with the limits on $a_x \sin i$ derived in the previous section.

Despite the above similarities with 4U1626–67, there are also important differences. First, the X–ray spectrum is much softer than that of 4U1626–67, . In this respect 4U0142+61 is more similar to 1E2259+586, whose spectrum can be described by a power law with energy index ~3, plus some possible cyclotron features suggesting a magnetic field B~5×10^{11}G (/12/). Second, the flux in the ultrasoft spectrum of 4U0142+61 is rather stable, unlike most accreting X–ray pulsars, and 4U1626–67 in particular which shows quasi–periodic flares on a timescale of 1000 s.

Even if the measured spin-down rate is the result of Doppler shifts due to the orbital motion, 4U 0142+61 must be close to being an equilibrium rotator, implying a neutron star magnetic field of B~10^{12}(d/4 kpc) G.

An Isolated Neutron Star ?

The possibility that 4U0142+61 is an isolated neutron star is suggested by its very high F_x/F_{opt}, its ultrasoft spectrum, the absence of significant variability on long timescales and the spin–down of the period (/7/). In principle the X–ray emission could be due to non–thermal magnetospheric processes powered by rotational energy, to thermal emission from the neutron star surface, or to accretion from the interstellar medium. While examples of the first two mechanisms are well known (/13/), no compelling evidence for a compact object accreting from the interstellar medium has yet been found, despite several studies show that such sources could be relatively common (/14/; /15/).

For a spin period of 8.7 s and any reasonable magnetic field ($\leq 10^{14}$ G), the available rotational energy loss is too small, unless 4U0142+61 is at a distance of a few parsecs. To investigate the possibility of thermal emission, we fitted blackbody spectra to the 1985 ME+LE data, obtaining kT$\sim$0.5 keV, $N_{II}\sim$2–4$\times10^{21}$ cm^{-2} and a bolometric luminosity $\sim10^{32}(d/100$ pc$)^2$ ergs s^{-1} (note however that these fits gave high reduced χ^2 of $\sim$3–6). This implies an emission region of $\sim$0.1$(d/100$ pc$)^2$ km^2 , compatible with a hot spot on the surface of a neutron star, possibly the magnetic polar cap heated by accretion from the interstellar medium. In this case the accretion induced luminosity would be $\sim10^{32}(v/50$ km s$^{-1})^{-3}(n/100$ cm$^{-3})$ ergs s^{-1}, where v is the neutron star velocity relative to the interstellar medium of density n (/15/). The already mentioned molecular cloud (/10/) could be at a distance of only a few hundred parsecs and easily provide the density required to power the observed luminosity. For accretion onto the neutron star to take place, the magnetospheric centrifugal barrier must be open, and the low rates implied by the above luminosity therefore require a magnetic field B$\leq10^{10}(d/100$ pc$)$ G (see, e.g. /16/).

CONCLUSION

The 8.7 s pulsations, together with the very high F_x/F_{opt} indicate that the compact object in 4U0142+61 is a rotating magnetised neutron star. The most probable model involves a LMXRB with a very faint companion, similar to 4U1626–67. 4U 0142+61 provides a counterexample to the phenomenological criterion that an ultrasoft X–ray spectral component is a signature of accreting black holes (/1/).

REFERENCES

1. N. E. White and F. E. Marshall, *ApJ*, 281, 354 (1984).

2. N. E. White et al., *MNRAS*, 226, 645, (1987).

3. C. Motch et al., *A&A*, 246, L24, (1991).

4. S. Mereghetti, L. Stella and F. De Nile, *A&A*, 278, L23, (1993).

5. G. L. Israel, S. Mereghetti and L. Stella, *IAU Circ.* n. 5889, (1993).

6. G. L. Israel, S. Mereghetti and L. Stella, *ApJ Lett.*, 433, L25.

7. C. Hellier, *IAU Circ.* n. 5994, (1994).

8. S. R. Davies et al., *MNRAS*, 245, 268, (1990).

9. R. H. D. Corbet and C. S. R. Day, *MNRAS*, 243, 553, (1990).

10. D. Leisawitz, F. N. Bash and P. Thaddeus, *ApJ Supp.*, 70, 731, (1989).

11. F. Verbunt, R. A. M. J. Wijers and H. M. G. Burm, *A&A*, 234, 195, (1990).

12. K. Iwasawa, K. Koyama and J. P. Halpern, *PASJ*, 44, 9 , (1992).

13. S. Mereghetti, P. Caraveo and G. F. Bignami, *ApJ Supp.*, in press.

14. A. Treves and M. Colpi, *A&A*, 241, 107, (1991).

15. O. Blaes and P. Madau, *ApJ*, 403, 690, (1993).

16. L. Stella, S. Campana, M. Colpi, S. Mereghetti and M. Tavani, *ApJ*, 423, L47, (1994).

PII: S0273-1177(97)00036-7

Adv. Space Res. Vol. 19, No. 1, pp. (1)45–(1)54, 1997
© 1997 COSPAR
Printed in Great Britain. All rights reserved
0273–1177/97 $17.00 + 0.00

SPECTRAL DISTINCTION BETWEEN BLACK HOLES AND NEUTRON STARS: THE CONTRIBUTION OF THE SIGMA TELESCOPE

P. Laurent and M. Denis

Service d'Astrophysique, C.E.N. Saclay, 91191 Gif-sur-Yvette Cedex, France

ABSTRACT

After more than four years of operation, the imaging γ−ray telescope SIGMA has accumulated several days of observation toward black holes and neutron stars. So far, more than twenty five of them have been detected with SIGMA. Although all these sources have spectra dominated by comptonisation, it is remarkable that suspected black hole systems are particularly numerous among the most luminous sources above 150 keV. Also, two black hole candidates have shown a strong transient emission around 511 keV, which has been interpreted as electron−positron annihilation features. We will discuss in this paper the reliability of these two phenomena, a strong luminosity above 150 keV and the emission of an annihilation line, as potential black hole criteria.

BLACK HOLE CRITERIA

As they have quite similar mass and compactness, a black hole and a non-magnetized neutron star in a binary system produce nearly the same effects to their environment. They are then very difficult to distinguish observationally, and the only sure criterion we have to discern between them is their mass. So, as a neutron star mass theoretically cannot exceed ~ 3 $M_\odot$, any object which have a greater measured mass is declared a "black hole candidate".

With a most probable mass of 7 $M_\odot$ /1/, Cygnus X−1 was the first genuine black hole candidate. Accordingly, all its properties have been investigated as possible black hole criteria (see Tanaka /2/ for a review). The best X−ray spectral criterion seems now to be the observation of a single power law down to keV energies when the overall X−ray luminosity is stronger than 10^{37} erg/s. A neutron star indeed might show a black body component with kT $\approx$ 2 keV due to its heating by the accreted material /2/.

In the same way, as the SIGMA energy range (30 keV − 1 MeV) is devoted mostly to the observation of black holes and neutron stars (see Table 1), we have tried to define other criteria using SIGMA data. The study of the spectra of these sources has revealed us two phenomena which could be regarded as signatures of black holes: firstly, all the most luminous SIGMA sources above 150 keV are black hole candidates, and then observed transient emission of annihilation lines come from two sources which are also two good black hole candidates. In this paper, we will explore in more details this two possible criteria and see how far they are reliable.

P. Laurent and M. Denis

TABLE 1 SIGMA catalog (6 January 1994).

SIGMA NAME	OTHER NAMES	BEHAVIOR	NATURE
SIG 0418+327	Nova Persei	Exponential Decay	Stellar Black Hole ?
SIG 0531+219	M1/PSR 0531+21	Stable, periodic	Nebula/ radio pulsar
SIG 0834−429	GS 0834−429	Transient	Neutron Star
SIG 1124−683	Nova Muscae	Exponential Decay	Stellar Black Hole
SIG 1208+396	NGC 4151	Slightly variable	Seyfert 1
SIG 1223+129	NGC 4388	Slightly variable	Seyfert 2
SIG 1226+023	3C 273	Slightly variable	Quasar
SIG 1227+025	GRS 1227+025	Transient	Quasar z=0.57 ?
SIG 1322−428	Centaurus A	Variable	AGN
SIG 1510−590	PSR 1509−58	Stable	Nebula?/radio pulsar
SIG 1524−616	TrA X-1	Transient	Stellar Black Hole ?
SIG 1656−415	OAO 1657−415	Transient, periodic	Neutron Star
SIG 1659−487	GX 339−4	Very variable	Stellar Black Hole ?
SIG 1700−377	4U 1700−37	Very variable, eruptive	Neutron Star ?
SIG 1716−249		Transient	Stellar Black Hole ?
SIG 1724−307	Terzan 2	Persistent	Neutron Star ?
SIG 1728−338	4U/MXB 1728−34	Very variable, eruptive	Neutron Star
SIG 1729−247	GX 1+4	Very variable, periodic	Neutron Star
SIG 1731−260	KS 1731−260	Transient	Neutron Star ?
SIG 1732−304	Terzan 1	Transient	Neutron Star ?
SIG 1734−292	GRS 1734−292	Transient	Neutron Star ?
SIG 1734−269	SLX 1735−269	Persistent	
SIG 1740−297	1E 1740.7−2942	Very variable	Stellar Black Hole ?
SIG 1742−288	GRS 1741.9−2853		
SIG 1742−294	A 1742−294	Transient	Neutron Star
SIG 1743−290		Variable	
SIG 1747−341		Persistent	
SIG 1758−257	GRS 1758−258	Very variable	Stellar Black Hole ?
SIG 1912+108	GRS 1915+105	Transient	
SIG 1956+350	Cygnus X-1	Variable	Stellar Black Hole

HIGH LUMINOSITY ABOVE 150 keV ?

Table 2 shows that indeed black hole candidates are intrinsically more luminous sources than neutron stars above 150 keV. The luminosity of black holes above 150 keV deduced from the SIGMA observations is greater than 10^{36} erg s^{-1}, to be compared to less than 2×10^{35} erg s^{-1} for neutron stars. So, the $L_{>150}$ criterion seems reliable. We can notice then that this criterion does not depend on the strength of the neutron star magnetic field, as pulsars as well as bursters are present in Table 2.

TABLE 2 Black holes vs neutron stars comparison (May 1993). Question marks means that the source distance is poorly known

	Sources	$L_{150-500}$ (10^{35} erg/s)
Black Hole Candidates	Cygnus X−1	≈ 20
	GX 339−4	≈ 30
	Nova Muscae	≈ 50
	Nova Persei	≈ 115 ?
	GRS 1758−258	≈ 20
	1E 1740.7−2942	≈ 40
Neutron Star	4U 1700−377	≈ 5
	OAO 1657−415	≤ 1
	A 1742−294	≤ 5
	GX 1+4	≤ 2
	KS 1731−260	≤ 2 ?
	GX 354−0	≤ 3
	Crab	≈ 40

In fact, this criterion does not depend also on the underlying emission process, which could be thermal or non-thermal. However, the more commonly accepted explanation, i.e. the "Compton cooling", for this phenomena is a thermal one /3/. It suppose that the hard X−ray emitting plasma around neutron stars or black holes is cooled by soft photons (Compton scattering) and warmed by the accreted protons (Coulombian collisions). The disk and an eventual corona are natural sources of cooling soft photons for the two types of object. But a neutron star solid surface acts as an additional source of cooling photons, in such a way that the temperature of the medium close to the compact object has to be lower in neutron star systems. It is this difference that we will try to precise below, deriving a "mean" electron temperature for the hard X−ray emitting medium in the two type of sources.

Compton Cooling In Black Hole Candidates

As shown in Figure 1, black hole candidates spectra are generally well-fitted with a comptonized Sunyaev-Titarchuk model (hereafter ST model /4/). Table 3 gives the result of this fitting for all the candidates we have detected with SIGMA. We see in this table that the electron temperature of these black hole candidates is always higher than 30 keV.

It can be seen from Table 3 that, for an unknown reason, novae have generally a higher temperature than persistent sources. We can note also that among these persistent sources, the three which are supposed to be in low mass systems (GX 339−4, 1E 1740.7 −2942, and GRS 1758 −258) have a significantly lower temperature than the high mass system Cygnus X−1. This may be an indication of a possible correlation between the electron temperature and the companion type.

TABLE 3 Spectral fits of black hole candidates with a Sunyaev-Titarchuk model

SOURCE NAMES	F_{100} (10^{-4} ph/cm²/s/keV)	kT_e (keV)	τ_{Disk}
Nova Muscae	4.4 ± 0.1	74 ± 30*	0.4 ± 0.2
Nova Persei	11 ± 0.3	58 ± 1	2.0 ± 0.05
Nova Ophiuchi	6 ± 0.1	45 ± 2	0.75 ± 0.03
GX 339−4	1.8 ± 0.1**	34 ± 6	1.3 ± 0.4
1E 1740.7 −2942	0.7 ± .06***	38 ± 12	1.5 ± 0.5
GRS 1758 −258	0.5 ± 0.02	35 ± 5	1.7 ± 0.8
Cygnus X−1	6.4 ± 0.1	63 ± 6	2.1 ± 0.2

    * Best fitted by a power law of index ≈ -2.3
    ** "High state"
    *** "Standard state".

Moreover, if we look closely to the spectrum of Nova Persei shown in Figure 1.b, we see that the points deviate significantly from the fit above 300 keV. This behavior has also been observed by OSSE for Cygnus X−1 and GX 339−4 /5/. Three different non-exclusive explanations of this phenomenon have been given up to now:

1) Above some tenth of a keV, electrons are relativistic, and the ST model based upon the non-relativistic Thomson cross section is no longer valid. A relativistic treatment using the Klein-Nishina cross section can account for the high energy observed excess /6/.

2) It is also quite certain that it exists in the inner part of the disk several regions with different temperatures. So, a multi-temperatures model is somewhat more suitable /7/. For instance, The Cygnus X−1 spectrum can be well fitted by the superposition of two ST models /5/.

3) If the temperature in the inner region is high, pions production can occur, and the electromagnetic cascade resulting from the annihilation of these pions can give birth to an additional component similar to the observed excess /8/.

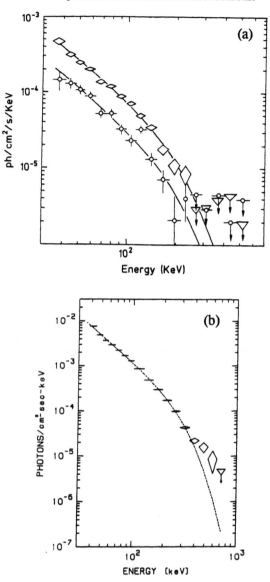

Fig.1 (a) Spectrum of 1E 1740.7-2942 in the "standard" and "sub luminous" states /9/; (b) Spectrum of Nova Persei /10/. The best fit with a ST model is also shown

All these models require a very high electron temperature ($\approx$ 100 keV) close to the black hole. So, the "mean" temperature we search can be of the order of one hundred keV. This is furthermore corroborated by the observation of annihilation lines in two black hole candidates. Let us see now what could be the value of this "mean" temperature in the vicinity of neutron stars.

P. Laurent and M. Denis

Compton Cooling In Neutron Stars

During the years 1990−1994, SIGMA had detected many neutron stars, mostly in binary systems (see Table 1). Among these binaries, only two of them were bright enough to allow detailed spectral fits: GX 1+4 and 4U 1700−377. GX 1+4 is a 2 minutes period binary X−ray pulsar close to the galactic center /11/, and 4U 1700-377 is a 3.4 day eclipsing massive system /12/. The 4U 1700−377 X-ray spectrum and the optical mass function show that the X-ray source should be a pulsar, although no X-ray pulsations have been observed so far /13/.

The observation-to-observation spectra of these sources are generally well fitted by a thermal bremsstrahlung model with a temperature of about 40 keV, which is comparable to the lower temperature obtained from black holes. We may think then that the cooling due to the soft photons produced at the surface of the heated neutron star is not so efficient. But, as this thermal bremsstrahlung fit gives us also the emission measure of the source, we can verify the implicit assumption of an optically thin emitting medium. Given a typical radius of the emitting region (10^8 cm, for the magnetosphere for instance), and the source distances we find that the Thomson optical depth of this medium is larger than 1. So, the hypothesis of a optically thin medium currently used in the past is not justified, or only for an emitting region of radius greater than 10^{10} cm. We have then to use comptonized models to describe the spectra. Using a ST model, we find the parameters given in Table 4 for the two sources under consideration.

TABLE 4 Spectral fits of GX 1+4 and 4U 1700−377 with a ST model

Fit parameters	GX 1+4 (1991)	4U 1700−377 (Normal state)	4u 1700−377 (Flare state)
F_{100}*	1.5 ± 0.3	0.35 ± 0.13	1.33 ± 0.34
kT_e	17 ± 7	14.3 [12,22]	15.3 [12,20]
τ**	5 ± 3	12.0 [3.5,+∞[	6.9 [3.9,+∞[

* Flux at 100 keV in 10^{-4} photon/cm²/s/keV
** Thomson optical depth (spherical cloud)

The temperatures we get by taking into account comptonisation processes are surprisingly stable, around 15 keV, while the flux can vary by a factor four. This temperature is now much lower than the one derived for the black holes candidates ($\approx$ 50 keV and probably around 100 keV in some systems), showing that the plasma surrounding neutron stars is significantly cooler.

THE PRESENCE OF AN ANNIHILATION LINE ?

Another possible black hole criterion which can be derive from the SIGMA observations is the presence of an annihilation line. But, whereas it is true that the lines we observed was emitted from black hole candidates, there is no reason *a priori* to suppose that this emission cannot occur also in neutron star binaries. Indeed, non thermal processes around a neutron star in a low mass system can theoretically produce pairs, and then give rise to a 511 keV line /14/. So, the observation of an annihilation line does not seem to be, for the moment, a good black hole criterion *per se*. Nevertheless, such an observation can give precise information on the geometry and the physics of the system in consideration, which can help for the determination of the nature of the compact object, as we will see below.

The Annihilation Line in 1E 1740.7−2942

1E 1740.7−2942 is one of the two hard X-ray sources seen above 150 keV in the galactic bulge. This source has been continuously monitored by SIGMA since 1990, showing strong intensity variations on month time scales /15/. Due to the similarity of its 35−300 keV spectrum with the Cygnus X−1 one, 1E 1740.7−2942 may be thought as a black hole candidate. On October 13−14, 1990, the source underwent a strong high energy outburst above 200 keV, whose shape suggests positron annihilation radiation /16,17/. The observation of this event has triggered a multi-wavelength campaign of observations of this source, which has revealed in the radio domain a double-sided jet and a central core which was attributed to the compact object /18/.

The annihilation line seen in 1E 1740.7−2942 has been modelled by Maciolek-Niedzwiecki & Zdziarski /19/. They have shown that the line must have been produced in a medium with a temperature of nearly 40 keV consistent with the temperature derived from the underlying continuum, and very close to the compact object (~ 3 R_s, where R_s is the Schwartzchild radius). This rules out the candidature of a less than 1.5 $M_\odot$ neutron star as the compact object, as this means that the line would be produced below its surface.

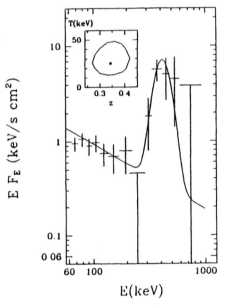

Fig. 2. Spectrum of 1740.7-2942 during the September 1990 event. The best fit with the Maciolek-Niedzwiecki & Zdziarski model is also shown /19/.

Also, as the line is not broadened by Compton scattering, the optical thinness of the medium enable the authors to derive some constraints on the geometry of the emitting region. After having ruled out spherical geometries, they show that the most likely emitting medium for the

511 keV photons is the jet, seen edge on. As we said above, this illustrates that the observation of an annihilation line can lead to get severe constraints on the geometry of the system. We will see now that this is even more true in the case of Nova Muscae.

The Nova Musca Case.

SIGMA was conveniently on hand when a nova appeared in January 1991. The GRANAT observatory has then monitored the source from the beginning of the flare /20,21/. The light curve in the SIGMA domain is very different from the one obtained at X−ray energies, showing a very sharp increase a few days before the soft X−ray peak, which then decayed very fast. Afterwards, the light curve shows a long period of slow decrease (1 month), with a drop at hard X−ray energy at the same time as an enhancement of the soft X−ray flux. Then, after 6 months of absence, the source was again detected by SIGMA. Measurements of the mass function of the Nova Musca binary system give a lower limit for the compact object mass of 3.1 ± 0.5 $M_\odot$ /22/, which put definitively this nova on the list of black hole candidates.

The source spectral shape observed has not varied much during the flare, and is well modelled by a power law of index $\approx$ -2.4. On January 20−21, 1991, a high energy feature appeared during nearly 10 hours in the spectrum (see Figure 3). It was centered at 480 keV with a line width (FWHM) lesser than 100 keV, and was accompanied by another feature at lower energy ($\approx$ 200 keV). If the 480 keV line is due to an e^+ e^- annihilation process, its width indicates an emitting region temperature lower than 10^8 K. The presence of another feature at lower energies, if due to the Compton backscattering, also induces, beside a high Thomson optical depth, a low temperature ($<$ 10 keV). Moreover, as it is clearly seen from Figure 3, the underlying spectrum keeps the same spectral shape from 10 to 400 keV, with no evidence of a low energy excess (around 100 keV) due to multiple Compton scattering. This rules out most spherical and disk geometry for the scattering medium /23,24/.

In fact, we have shown /24/ that the only way to take into account all these constraints is to consider that there is a jet in Nova Muscae. This seems quite reasonable as, up to now, three jets have been detected from black hole candidates (1E 1740.7−2942, GRS 1758−258, and Nova Ophiuchi). If, furthermore, we make the reasonable assumption that the end of the jet far from the compact object is relatively cool ($<$ 10 keV), a relatively narrow Doppler redshifted annihilation line, and a blueshifted backscattering line can be efficiently produced and detected if we look at a relativistic jet face on /24/. On the other hand, the backscattering line would be strongly broadened at higher inclinations due to multiple scatterings.

The narrowness of the observed backscattered line implies then that if there was effectively a jet in Nova Muscae, we have looked at it face on. So, as this implies that the inclination of the system is around 10° or lower, we can deduce from the mass function /22/ that the mass of the compact object is greater than 30 $M_\odot$. This rather high mass may be an argument against our jet model. If so, we don't know any other geometry which can account for the phenomenon observed in Nova Muscae /24/. So, we may have to reject one of the following hypothesis:

. The 480 keV line is due to electron-positron annihilation and/or
. the 200 keV line results from the backscattering of the 480 keV line.

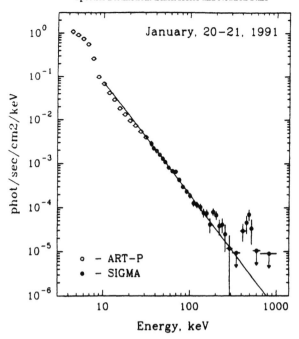

Fig. 3. Spectrum of Nova Muscae (January 1991) /21/.

CONCLUSION

We have seen that the $L_{>150}$ criterion seems reliable for the identification of black holes. If we suppose that this criterion results from the "Compton cooling" effect, SIGMA observations have shown that the "mean" temperature of the hard X−ray emitting plasma is indeed lower for neutron stars ($\approx$ 15 keV), than the one of black holes ($>$ 50 keV). In another way, the emission of annihilation lines can theoritically occured in neutron star and black hole systems. Nevertheless, we have shown that the observation of lines from compact objects, even if it is not a black hole criterion *per se*, gives strong indications on the nature of the central compact object.

REFERENCES

1. Gies D.R., Bolton C.T., 1986, ApJ, 304, 371.

2. Tanaka Y., 1989, Proc. 23rd ESLAB Symp., Italy, ESA SP-296, 1

3. Sunyaev R., Churazov E., Gilfanov M., et al., 1991, A&A 247, L29

4. Sunyaev R.A., Titarchuk L.G., 1980, A&A, 86, 121.

5. Johnson W.N., Kurfess S.D., Purcell W.R., et al., 1993, A&AS, 97, 21

6. Grebenev S., Sunyaev R., Pavlinski M., et al., 1993, A&AS, 97, 281

7. Liang E., 1991, Proc. Compton Observatory Science Workshop, Annapolis, NASA, 3137, 173

8. Jourdain E., Roques J.P., 1994, ApJ, 426, L11

9. Cordier B., Paul J., Goldwurm A., et al. 1993, A&A, 272, 277

10. Roques J.P., Bouchet L., Jourdain E., et al., 1994, ApJS, 92, 451

11. Laurent P., Salotti L., Paul J., et al., 1993, A&A, 278, 444

12. Laurent P., Goldwurm A., Lebrun F., et al., 1992, A&A 260, 237

13. Gottwald M., White N.E., and Stella L., 1986, MNRAS, 222, 21P

14. Kluzniak W., Ruderman M., Shaham J., and Tavani M., 1988, Nature, 336, 558

15. Cordier B., Paul J., Goldwurm A., et al., 1993, A&A, 272, 277

16. Bouchet L., Mandrou P., Roques J.P., et al., 1991, ApJ, 383, L45

17. Sunyaev R., Churazov E., Gilfanov M., et al., 1991, ApJ, 383, L49

18. Mirabel I.F., Rodriguez L.F., Cordier B., et al., 1992, Nature, 358, 215

19. Maciolek-Niedzwiecki A., Zdziarski A.A., 1994, to appear in ApJ

20. Goldwurm A., Ballet J., Cordier B., et al., 1992, ApJ, 389, L79

21. Sunyaev R. A., Churazov E., Gilfanov M., et al., 1992, ApJ, 389, L75

22. Remillard R.A., Mc Clintock J.E., Bailyn C.D., 1992, ApJ, 399, L145

23. Hua X-M, Lingenfelter R.E., 1993, ApJ, 416, L17

24. Laurent P., 1994, in preparation

 Pergamon

Adv. Space Res. Vol. 19, No. 1, pp. (1)55–(1)61, 1997
© 1997. Published by Elsevier Science Ltd on behalf of COSPAR
Printed in Great Britain. All rights reserved
0273–1177/97 $17.00 + 0.00

PII: S0273-1177(97)00037-9

LMXBS AND BLACK HOLE CANDIDATES IN THE GALACTIC CENTER REGION

E. Churazov*, M. Gilfanov*, R. Sunyaev*, B. Novickov*,
I. Chulkov*, V. Kovtunenko*, A. Sheikhet*, K. Sukhanov*,
A. Goldwurm**, B. Cordier**, J. Paul**, J. Ballet**,
E. Jourdain***, J. P. Roques***, L. Bouchet*** and
P. Mandrou***

** Space Research Institute, Russian Academy of Sciences, Profsouznaya
84/32, 117810 Moscow, Russia*
*** Service d'Astrophysique, Centre d'Etudes Nucleaires de Saclay, 91191
Gif-sur-Yvette Cedex, France*
**** Centre d'Etude Spatial des Rayonnements, 9, Avenue du Colonel Roche,
BP 4346, 31029 Toulouse Cedex, France*

ABSTRACT.

The central part of the Galactic Plane has been intensively observed by GRANAT/SIGMA in 1990-1994. At least 11 sources are seen on the averaged 35-100 keV image of this region, most of them being low mass X-ray binaries. With GRANAT/SIGMA observations the X-ray bursters are recognized as sources emitting hard X-rays (up to $\sim$ 100 keV) during substantial fraction of time. This "bimodal" behavior (soft and hard spectral states) of X-ray bursters resembles that of black hole candidates (BHC), although unlike BHCs soft state for the bursters seems to be the dominant one. On average the spectra of bursters in the SIGMA energy band (above 35 keV) were found to be steeper than that of BHCs. This difference is discussed in terms of Comptonization model.

INTRODUCTION

Part of the Galactic Plane adjacent to the Galactic Center has been observed by GRANAT/SIGMA in hard X-rays during more than 10^7 s in past five years. All images obtained during individual observations were added together and resulted map (in units of standard deviations) is shown in Fig.1. Since during most of the observations GRANAT was pointed towards the Galactic Center itself the sensitivity at $(l = 0, b = 0)$ is much higher that at the outer parts of the map. At least 11 sources are clearly seen on this map including four black hole candidates (GX339-4, 1E1740.7-2942, GRS1758-258, GRS1716-249), two X-ray pulsars (GX1+4, OAO1657-415), one nonpulsing HMXB (4U1700-37), several X-ray bursters (4U1728-337 (GX354-0), 4U1724-307 (Terzan 2), 4U1732-304 (Terzan 1)) and the source SLX1735-269 (neither pulsations nor X-ray bursts have been detected from it so far). A number of bursters, which are not seen on the map averaged over all observations (e.g. KS1731-260, A1742-294, 4U1705-44), have been detected only during short periods of time. With GRANAT/SIGMA observations X-ray bursters are recognized as the objects emitting hard X-rays (up to $\sim$ 100 keV) during substantial fraction of time.

SPECTRA OF X-RAY BURSTERS AND BLACK HOLE CANDIDATES

From SIGMA point of view (i.e. at hard X-rays) nonpulsing LMXBs can be roughly separated into three groups, which differ in the Spectral Energy Distributions (SED) in the energy range from few keV to hundreds of keV as shown in Fig.2.

High luminosity ($\sim 10^{38} erg/s$) Z-sources are very bright in standard X-ray band but they do not emit in SIGMA band more than $\sim$ few % of their luminosity (Fig.2a). On the contrary for black hole candidates (Fig.2c) bulk of the luminosity is emitted at the energies $\sim$ 100 keV during their hard state. The sources known as the X-ray bursters (from the list of SIGMA detections) constitute intermediate group (Fig.2b): typically their spectra are rather soft and roughly resemble Fig.2(a), while for some periods SEDs more resemble that of BHCs (Fig.2c), although they peak at

E. Churazov *et al.*

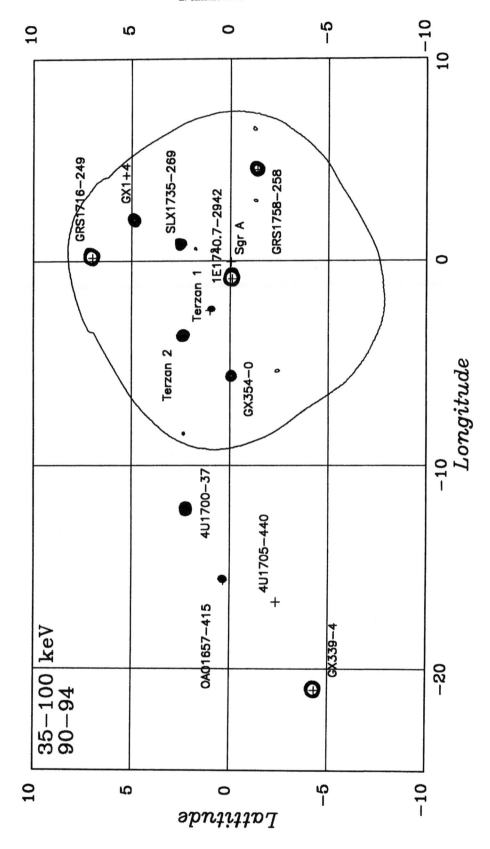

Figure 1: Combined image of the central part of the Galactic Plane in the 35-100 keV band averaged over all SIGMA observations in 1990-1994. Contours are in standard deviations starting from 4.5σ with 0.5σ increment. Sensitivity is highest around the GC. Thin line show the area with sensitivity 0.5 of the maximal value.

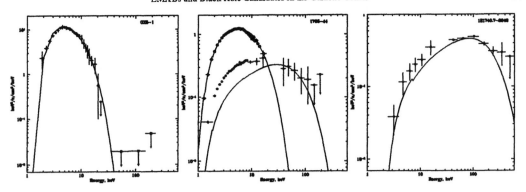

Figure 2: Spectral energy distribution for GX5-1 (left), 4U1705-44 (middle) and 1E1740.7-2942 (right). Points above 35 keV are SIGMA data (averaged over 1990-1994 for GX5-1 and 1E1740.7-2942 and averaged over 1994 Spring for 4U1705-44). Observations in standard X-ray band for GX5-1 and 1E1740.7-2942 are from TTM /1/, for 4U1705-44 from EXOSAT ME. For comparison the SEDs for the bremsstrahlung model with low energy absorption are shown with temperature ∼5 keV (left), 5 and 40 keV (middle), 130 keV (right).

somewhat lower energies ∼ 30-50 keV. The latter conclusion is based on the assumption that hard states of the X-ray bursters observed in standard X-ray band (e.g. 4U1705-44 /2/; 4U1608-52 /3/, 4U1732-304 (see Fig.5)) correspond to the hard states observed by SIGMA above 35 keV. This is not necessarily true because of the lack of simultaneous broad band observations.

Using SIGMA observations of the Galactic Center region it is possible to compare typical spectra of X-ray bursters and BHCs above 35 keV. Shown in Fig.3 are the parameters of approximation of their spectra in the 35-150 keV band by the power law and thermal bremsstrahlung models. Note that for simplicity the spectra have been accumulated ignoring the slight variations of the detector energy response with time and more rigorous treatment of calibration for individual sources may lead to slightly different values of photon index and temperature. One can see that both in terms of power low and bremsstrahlung approximations X-ray bursters have steeper spectra in the SIGMA band than BHCs. No one of known X-ray bursters is located close to the group of BHCs on this diagram. The only exception from this rule is the hard (photon index ∼ 1.7) spectrum of 4U1724-307 (Terzan 2) observed by SIGMA in 1990 Spring /4/. During later observation of this source it's spectrum was steeper (Fig.3). Because the 4U1724-307 position coincides with the globular cluster Terzan 2 contamination from another source during 1990 Spring observations can not be ruled out. Alternatively it might mean that conclusion that X-ray bursters have photon index steeper than ∼2.5 in the 35-150 keV band is true only on average.

COMPARISON OF THE SPECTRA IN TERMS OF COMPTONIZATION MODEL

Comptonization of the low frequency photons in the cloud of hot electrons is considered as the plausible mechanism of the formation of hard components in the spectra of accreting binary sources. As is known there is close relation between the slope of the emergent spectrum and the luminosity enhancement factor (see e.g. /5,6,7/). Enhancement factor A characterizes the ratio of the energy flux of comptonized spectrum to the initial energy flux of the soft photons. If the comptonization is the only cooling mechanism for the hot electrons then $A - 1$ equals to the ratio of energy release in the cloud to the energy flux of the soft photons. Using modified Kompaneetz equation (i.e. uniform media with constant probability of escape) one can express enhancement factor as $A^{-1} = 1 - y \times (1 - \frac{<E^4>}{<E^3>}/4T_e)$, where $y = 4\frac{kT_e}{m_e c^2} \cdot max(\tau, \tau^2)$ is the comptonization parameter and $\frac{<E^4>}{<E^3>} = \int E^4 n dE / \int E^3 n dE$ is the intensity-weighted mean energy /5/. For steep enough spectra $\frac{<E^4>}{<E^3>}/4T_e \ll 1$ and $A^{-1} = 1 - y$. Photon index in turn also depends on y: $\Gamma = -1/2 + (9/4 + 4/y)^{1/2}$. The greater is the enhancement factor the harder is the comptonized spectrum. Convenient approximate relation between enhancement factor and photon index in the nonrelativistic limit is given in /7/: $A = 1 + \frac{1}{\Gamma-2} - \frac{1}{\Gamma-2}(h\nu_0/3T_e)^{\Gamma-2}$, where $h\nu_0$ is the initial energy of the soft photons. Shown in Fig.4a is the dependence of the photon index on the enhancement factor for the spherical

E. Churazov *et al.*

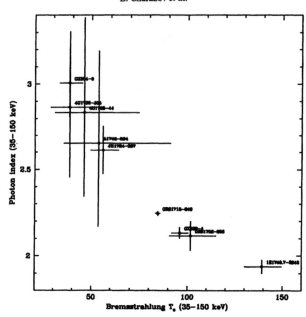

Figure 3: Parameters of power law and thermal bremsstrahlung approximations of the BHCs and X-ray bursters spectra in the 35-150 keV energy band. The data used: 1E1740.7-2942, GRS1758-258, GX354-0, 4U1732-304, 4U1724-307 (averaged over 1990-1994 observations), A1742-294 (1992 Fall), GRS1716-249 (1993 Fall), GX339-4 (1991 Fall), 4U1705-44 (1994 Spring). Sources known as X-ray bursters have systematically softer spectra in this band. Note that slight variations of detector energy response with time have been ignored while generating the data for this plot and more accurate analysis of the data for individual source may yield slightly different values, but the conclusion will remain the same.

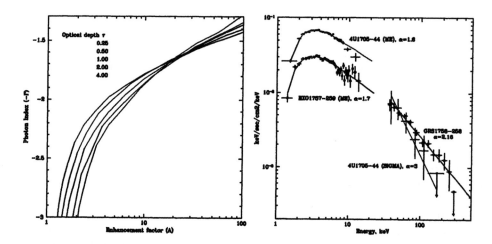

Figure 4: Dependence of photon index on the enhancement factor for the spherical geometry (left). Each curve corresponds to the fixed optical depth. Hard spectrum (photon index $\lesssim$ 1.7) in the standard X-ray band corresponds to the enhancement factor higher than ~10. Spectra of X-ray burster 4U1705-44 in the "hard" state and BHC GRS1758-258 (right). Low energy data – EXOSAT ME observations.

cloud with isotropic source of soft photons in the center. The Monte-Carlo code following /8/ has been used for these calculations. The initial energy of the soft photons has been fixed at 0.1 keV. The divergency of values of photon index at low enhancement factors is caused to some extent by the method of estimation of photon index. From Fig.4a it is clear that in order to provide hard spectrum with photon index $\lesssim 1.7$ enhancement factor must be greater than ~ 10. Thus hard spectra with photon index $\sim 1.4 - 1.7$ like observed in the standard X-ray band from black hole candidates indicate that enhancement factor is very high for these sources. Similar conclusion should remain valid for some other simple geometries of the problem when the soft photons source is embedded into the hot region having roughly spherical shape.

The comptonized spectra can be roughly characterized by two quantaties: position of the exponential cutoff and photon index of the spectrum at lower energies. The latter quantity is primarily the function of the enhancement factor. It is interesting to compare the spectra of neutron star binaries and BHCs observed by SIGMA from this point of view. It was suggested that because of the presence of the neutron star surface significant fraction of comptonized photons will be intercepted by the neutron star and returned back to the hot cloud in the form of soft photons, thus reducing the temperature in the hot region (see e.g. calculations of /9/). These agruments can be formulated in terms of enhancement factor. In particular in the slab geometry (hot slab above reprocessing surface – the geometry considered in /9/) about half of the comptonized radiation will be returned back to the soft source and reemitted to the hot region, thus restricting enhancement factor to the value $\sim 2-3$ (see e.g. /10, 11/). As the result one can expect (see Fig.4a) that shape of the spectrum will be steep both in hard and standard X-ray bands. While this seems to be in qualitative agreement with the slope of LMXBs spectra in SIGMA energy band (Fig.3), in standard X-ray band at least for several bursters (e.g. 4U1705-44, 4U1608-52, Terzan 1) the slope of the spectrum was found to be very hard (photon index ~ 1.6–1.7) during a fraction of time. Shown in Fig.4b are the spectra of X-ray burster 4U1705-44 and BHC GRS1758-258 (EXO1757-259) observed by EXOSAT (ME) along with spectra of these sources detected by SIGMA. The power law (with low energy absorption) fit to these data below ~ 10 keV gives photon index ~ 1.7 and ~ 1.6 for GRS1758-258 and 4U1705-44 respectively. This means that (in terms of simplified comptonization model discussed above) the enhancement factor was high ($A \gtrsim 10$) for both sources during these observations. Assuming further that hard states in standard X-ray band correspond to the hard states observed by SIGMA one can conclude that steepness of X-ray burster's spectra in SIGMA band is not due to steep power law but rather due to earlier cutoff of the spectrum.

The observations of hard power law component in the spectra of X-ray bursters mean that at least during their hard states there is no strong feed-back between hot region and reprocessing surface (i.e. flux of the soft photons entering hot region is not automatically ajusted to the level comparable with the energy release in the hot regions). This imposes some constraints on the geometry of the region. In particulary it might hint that hot clouds have the geometrical size much smaller then the distance to the reprocessing surface /11, 12/ or that there is sufficient Thomson optical depth between some of the hot regions and reprocessing surface. Note that optically thick accretion disk can itself serve as reprocessing surface /12/ so for the BHCs similar conditions may be required.

As soon as there is no strong feed-back between hot clouds and reprocessing surface hard spectrum can be formed due to the variability of the energy release in the hot clouds. Indeed if the energy in the clouds is released in the form of short flares then during these flares enhancement factor will be high and outgoing spectrum will be hard even if on the long time scales flux of soft photons exceeds the energy release in the hot clouds. This opens the possibility to form hard spectrum even if the clouds are illuminated by powerful (but stable) soft radiation (e.g. formation of hard tail during the "high" (soft) states of BHCs). Therefore in the absence of strong feed-back variability can be accompanied by hard tails (as observed for many sources). The stronger the variability is the harder is the outgoing spectrum. Note that for slab geometry and moderate values of optical depth of the hot region (i.e. strong feed-back) very hard spectrum can not be formed even if the variability is strong. Note further that as soon as enhancement factor is high (i.e. spectrum is hard) the slope of spectrum (below T_e) only weakly depends on the further increase of the enhancement factor.

E. Churazov *et al.*

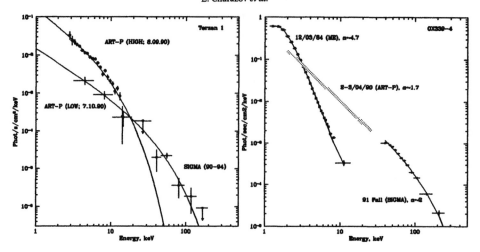

Figure 5: "Hard" and "soft" states of X-ray burster in Terzan 1 (left) and BHC GX339-4 (right). Low energy data: Terzan 1 (ART-P, /13/); GX339-4 (EXOSAT ME – soft state; ART-P – hard state /14/.

DISCUSSION

SIGMA observations above 35 keV have demonstrated that X-ray bursters spend substantial fraction of time in the "hard" states when they emit X-rays up to ~ 100 keV. It is likely that these hard states correspond to the hard states observed for a number of bursters in standard X-ray band (e.g. 4U1705-44, Terzan 1 etc.), although there are no simultaneous broad band observations of hard states of the bursters. If this is true then on the basis of SIGMA observations one can expect that sources most frequently appearing in SIGMA band can be easily found in hard state by the instruments operating in standard X-ray band (e.g. ASCA and XTE). Observations of these sources with XTE will be of prime importance because of the broad energy band. We can specially mention two sources: 4U1724-307 (Terzan 2) and 4U1728-337 (GX354-0). The former source is the best candidate since it was almost always present in the 35-100 keV band.

X-ray bursters seem to have two distinct states differing in the hardness of spectra. For many sources observed by SIGMA in the GC region correlation between the level of hard X-ray flux on the time scales of days has been observed (detailed analysis of these data will be published elsewhere). Since these time scales exceed by many orders of magnitude the characteristic time scales of the inner part of the accretion disk, hard states should be related to some parameter changing on these long time scales. The accretion rate is (of course) the first candidate to be considered. On the basis of HEAO-1 data it was found /15/ that weaker (lower luminosity) LMXBs have on average harder X-ray spectra. This trend was found comparing the sources which luminosities differ by almost three orders of magnitude. Using the bremsstrahlung fit to SIGMA data for calculation of the lower limit on the luminosity in the 2-150 keV band one can estimate that for A1742-294, 4U1705-44 and Terzan-1 the luminosity in the hard state is somewhat (factor ~ 2–4) lower than the 2-20 keV luminosity for the lowest "soft" states observed for these sources. On the other hand this luminosity is substantially lower than that for the highest "soft" states. These results suggest presence of "critical" accretion rate, below which changes in the accretion disk structure occur (see e.g. /16/).

These behavior of X-ray bursters looks similar to the bimodal behavior of BHCs, although the detailed shapes of the broad band spectra are different (Fig.5). One can assume that soft (dominant) states of bursters correspond to the soft (rarely observed) states of BHCs, appearance of which is defined by the accretion rate. This difference in the frequency of occurrence of soft states may indicate that actual parameter defining the disk structure is accretion rate in the units of critical (Eddington) accretion rate. Thus for similar luminosities (accretion rates) bursters will have systematically higher L/L_{cr} than that for BHCs.

This research has made use of data obtained through the High Energy Astrophysics Science Archive Research Center Online Service, provided by the NASA-Goddard Space Flight Center.

This works was supported in part by the Grant M4W000 from the International Science Foundation.

REFERENCES

1. Borozdin K. et al., 1994, Cospar, Report E1.2-040

2. Langmeier A. et al. 1987, *Ap. J.,* **323**, 288

3. Gottwald M. et al. 1987, *M.N.R.A.S.,* **229**, 395

4. Barret D. et al. 1991, *Ap. J. (Letters).,* **379**, L21

5. Shapiro S.L., Lightman A.P., Eardley D.M. 1976, *Ap. J.,* **207**, 187

6. Sunyaev R. & Titarchuk L. 1980, *A&A,* **86**, 121

7. Dermer C.D., Liang E.P., Canfield E. 1991, *Ap. J.,* **369**, 410

8. Pozdnyakov L., Sobol' I., Sunyaev R., 1983, Ap. Space Sc. Rev., Vol.2., p.189

9. Sunyaev R. & Titarchuk L. 1989, *in Proceedings of* "23rd ESLAB Symposium", ESA SP-296, Bologna, Italy, editors: J.Hunt & B.Battrick, v.1, p.627

10. Haardt F., Maraschi L. 1993, *Ap. J.,* **413**, 507

11. Gilfanov M. et al., 1993, Proc. of the Nato ASI "Lives of the Neutron Stars", Kemer, Turkey, in press

12. Haardt F., Maraschi L., Ghisellini G., Ap.J. (Letters), 1994, submitted

13. Pavlinsky M. et al., 1994, this issue.

14. Grebenev S. et al. 1993, *A&A (Supplement Series),* **97**, 281

15. Van Paradijs J. & Van der Klis M. 1994, *A&A,* **281**, L17

16. Barret D. & Vedrenne G. 1994, *Ap. J. (Supplement Series),* **92**, 505

Pergamon

PII: S0273-1177(97)00038-0

Adv. Space Res. Vol. 19, No. 1, pp. (1)63–(1)70, 1997
© 1997 COSPAR
Printed in Great Britain. All rights reserved
0273–1177/97 $17.00 + 0.00

ASCA OBSERVATION OF THE GALACTIC CENTER

Shigeo Yamauchi*, Yoshitomo Maeda**, Katsuji Koyama**
and ASCA Team

* *College of Humanities and Social Sciences, Iwate University, 3-18-34,
Ueda, Morioka, Iwate 020, Japan*
** *Department of Physics, Faculty of Science, Kyoto University, Sakyo-ku,
Kyoto 606-01, Japan*

ABSTRACT

X-ray imaging and spectroscopic observations near the Galactic Center in the 0.5–10 keV energy band were carried out with the X-ray satellite ASCA. Several X-ray sources were detected. Luminosities of the sources are estimated to be of order 10^{35}–10^{36} erg s^{-1}. The largely extended emission with K-shell transition lines from highly ionized atoms were confirmed. Moreover, an extended emission of the 6.4 keV line from the low ionized iron atom was discovered. The 6.4 keV line intensity is elongated to the north-east from the Galactic Center along the galactic plane and well correlates with the distribution of the dense molecular clouds observed in the radio bands. We interpret that the 6.4 keV line originates from the florescent from the molecular cloud irradiated by bright X-ray source.

INTRODUCTION

In the optical window, the Galactic Center has been obscured by an enormous amount of dust in the Galactic disk. This "dust curtain" has been removed by the opening of new windows in wave bands other than the optical. Radio and infrared observations have demonstrated that the Galactic Center is very bright and complex. Several expanding molecular clouds (or rings) suggest intermittent explosions near the Galactic Center, while the infrared observations reveal high concentration of stellar objects toward the Galactic Center /1/. Moreover, the gas motions observed in the radio and the infrared bands suggest large mass concentration at the central region /1/.

X-ray and gamma-ray bands can observe high energy phenomena occurred in the Galactic Center region. X-ray observations revealed the presence of several point sources and unresolved diffuse emission /2, 3, 4/. X-ray emission from the position of Sgr A* have been observed since its first detection by Einstein and reported that the flux is time variable /2, 3, 4, 5/. The maximum luminosity previously observed is 1.5×10^{36} erg s^{-1} /5/ which is much less than those of active galactic nuclei. A remarkable enhancement of the 6.7 keV emission line toward the Galactic Center would be the first X-ray indication of activity in the Galactic Center region, because the 6.7 keV line is attributable to Helium-like iron in an optically thin hot plasma with a temperature of 10^7–10^8 K /6, 7/. The total thermal energy and the dynamical time scale suggest that energetic explosion(s) of about 10^{54} erg took place in the Galactic Center region during the last 10^5 years. On the other hand, the detection of gamma-ray line emission from ^{26}Al also indicates that the occurrence of energetic phenomena, such as large numbers of supernovae or novae, near the Galactic Center within the last 10^6 years /8/.

Although many interesting features have been reported, what the activity of the central region of our Galaxy is and whether a massive black hole is located at the Galactic Center or not are still unsolved problems.

The ASCA satellite with its high spatial and spectral resolving power in the wide energy band makes a breakthrough. In this paper, we present preliminary results on the Galactic Center observed with ASCA. Throughout this paper, we assume that the distance of the Galactic center is 8.5 kpc.

OBSERVATIONS AND RESULTS

Observations of the Galactic Center region were carried out with both the Gas Imaging Spectrometer (GIS) and the Solid-state Imaging Spectrometer (SIS: X-ray CCD camera) on board ASCA on 1993 September 30 – 1993 October 7. The incident X-rays are collected by the X-ray Telescope (XRT) and detected by the SIS and the GIS placed on the focal plane. The field of view of the GIS is a circle of about 50' diameter, while that of the SIS is 22'×22' square. The set of X-ray detectors is sensitive in the 0.5–10 keV energy range. Details of ASCA and its instruments are given in ref. /9/.

(1) X-RAY SOURCES

Figure 1 displays the mosaic image of about 1°×1° region at the Galactic Center obtained with the GIS in the 0.7–10 keV band. We note that no correction of the non-uniform efficiency over the detector field was made. In this figure, several bright spots are clearly found. Three bright spots are identified to cataloged X-ray sources, 1E1743.1–2843, A1742–294, and 1E1740.7–2942. 1E1743.1–2843 and A1742–294 are considered to be X-ray binaries consisting of neutron stars and low mass normal stars, while 1E1740.7–2942 is famous for the positron-electron annihilation line /10/.

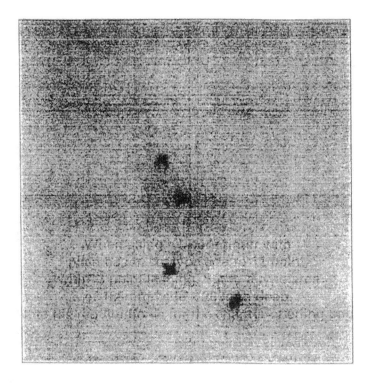

Fig. 1. A mosaic image of about 1°×1° region at the Galactic Center observed with the GIS detectors. The horizontal axis shows Right Ascension (1950.0), while the vertical axis shows Declination (1950.0).

Figure 2 is images of the central region obtained with the SIS in the 0.5–4 keV and the 4–10 keV bands. Two X-ray sources are clearly seen near the Galactic Center. In the soft X-ray band, diffuse-like X-ray emission is found to be around the position of Sgr A*. However, the hard band image, a point-like source located near the Sgr A* is bright. Hereafter, we refer to the source found in the soft band as a "soft source" and the source clearly seen in the hard band as a "hard source". The peak position of the soft source is consistent with that of Sgr A*. On the other hand, the position of the hard source is about 1 arcmin apart from the position of Sgr A* and is corresponding to that of the transient X-ray source, A1742–289, discovered by Ariel-5 within the 90% confidence level /11, 12/. Therefore, it is concluded that the hard source is not a galactic nucleus itself but a kind of galactic X-ray sources. The position of the X-ray sources are listed in table 1.

In order to investigate the nature of the X-ray sources, we extracted X-ray spectra and carried out the spectral fitting. The best-fit spectral parameters and X-ray luminosities of the sources are listed in table 1. From these parameters the X-ray luminosity is derived to be of order 10^{35}–10^{36} erg s^{-1}.

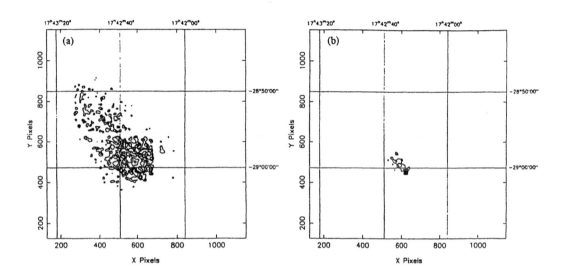

Fig. 2. Images of 20'×20' region at the Galactic Center observed with the SIS detectors in the 0.5–4 keV (a) and the 4–10 keV (b).

TABLE 1 Properties of X-ray sources

Source name	Position (RA, Dec)1950	Model	α kT (keV)	N_H ($\times 10^{23}$ Hcm^{-2})	L_X (2–10 keV) ($\times 10^{36}$ erg s^{-1})
1E1743.1–2843	(265.7908, -28.6993)	power-law	1.7±0.1	1.7±0.1	~ 2
Soft source	(265.6225, -28.9871)	thin thermal*	1.0±0.1	0.8±0.1	~ 2
			7.4±0.7	1.7±0.1	
Hard source	(265.6075, -29.0048)	power-law	2 (fixed)	4±2	~ 0.8
A1742–294	(265.7952, -28.7124)	power-law	1.5±0.1	0.6±0.1	~ 3
1E1740.7–2942	(265.1770, -29.7215)	power-law	1.2±0.1	1.2±0.1	~ 3

Note: * 2-temperature model.

In addition to these bright spots, a largely extended emission was detected all over the observed area. This is considered to be diffuse emission of 1°×2° size discovered by the 6.7 keV iron line mapping observations /6, 7/. Since the size of the diffuse component is much larger than the observed area, the outline of the component is not seen. Several K-shell transition lines from He-like atoms (1.87, 2.46,

3.13, and 6.7 keV lines) were found in the spectra, which strongly supports that the diffuse X-rays are due to an optically thin hot plasma with a temperature of $10^7 \sim 10^8$ K.

(2) DIFFUSE IRON LINE EMISSIONS

Figure 3 shows X-ray spectra of 4 positions around the Galactic Center. Emission lines at 1.87, 2.46, 3.13, 6.4 and 6.7 keV are found in the spectra. These lines except for 6.4 keV are identified to be K-shell resonance transition from He-like Si, S, Ar, and Fe within the systematic error of energy calibration of about 30 eV. These lines are probably attributable to thin hot plasma gas located around the Galactic Center /6, 7/. However, the 6.4 keV line is a K-shell transition line emitted from low ionized (cold) iron atom and is hardly understood in terms of the thin hot plasma.

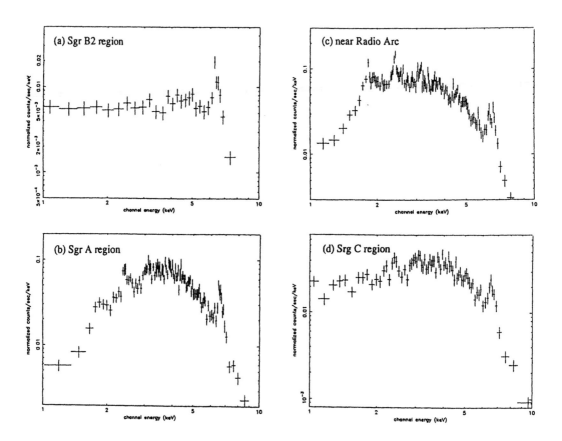

Fig. 3. X-ray spectra near the Galactic Center.

In order to investigate the global structure of the iron emission lines, we made narrow energy band map of the center energy of 6.4 keV and 6.7 keV. Figure 4 shows the line intensity map. The intensity distribution of 6.7 keV line shows a symmetrical shape roughly centered at the position of Sgr A*. Since the 6.7 keV line is attributable to an optically thin hot plasma of a temperature about 10^8 K, it follows a distribution of the high temperature plasma gas. On the other hand, the 6.4 keV line distribution is not symmetry with respect to the Galactic Center but elongated to the north-east from the Galactic Center along the galactic plane. Moreover, bright spots of the 6.4 keV line intensity are found around $l \sim 0°.7$ and $l \sim 0°.2$ which are corresponding to the positions of Sgr B2 cloud and the molecular cloud near "Radio Arc" as shown in figure 5.

(a) 6.4 keV line   (b) 6.7 keV line

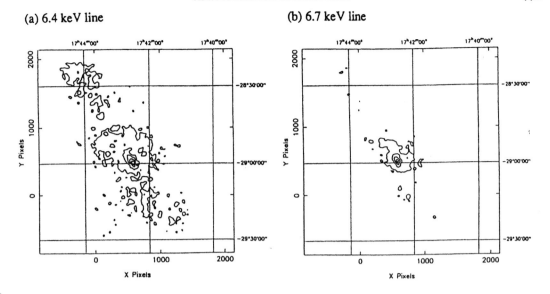

Fig. 4. Intensity maps of narrow energy band including line emissions from low ionized Fe (6.4 keV), and He-like Fe (6.7 keV).

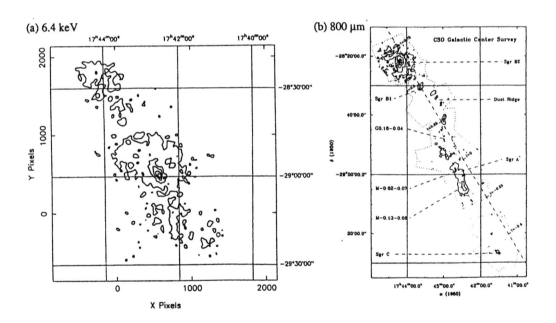

Fig. 5. Distributions of the 6.4 keV line (a) and the molecular cloud near the Galactic Center /13/ (b).

DISCUSSION

The most interesting discovery of the present observations is the 6.4 keV emission line. As mentioned above, the spatial distribution is different from that of the 6.7 keV line. This means that emission mechanism of the 6.4 keV line is different from that of the 6.7 keV (thin thermal emission from a high temperature plasma gas). On the other hand, the 6.4 keV line distribution well correlates with those of the molecular clouds observed in the radio band as shown in figure 5. What is the origin of the 6.4 keV line emission?

A strong 6.4 keV line is found in the spectrum of the Sgr B2 region shown in figure 6. The type II Seyfert galaxy NGC1068 also shows similar spectrum (figure 6, /14, 15/). NGC 1068 is considered that central X-ray source is hidden and only scattered X-rays can be observed. From the spectral fitting of the Sgr B2 region, the equivalent width of iron line and N_H value are determined to be about 2 keV and 10^{23} Hcm^{-2}, respectively. The equivalent width of iron line is nearly equal to that of NGC1068 /15/. From the spectral similarity, we can interpret that the strong 6.4 keV line is also due to a fluorescent by a bright X-ray source. Where is the bright X-ray source irradiating the cloud? This is the next question.

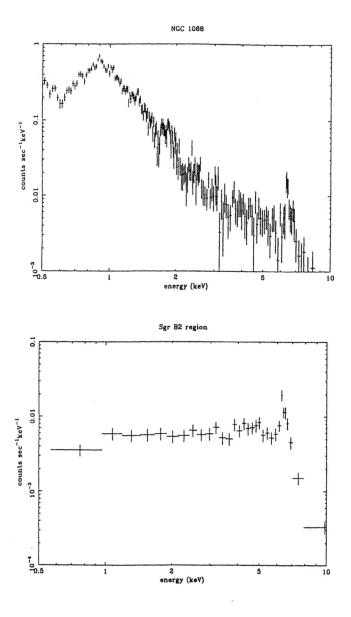

Fig. 6. X-ray spectra of NGC 1068 /15/ and the Sgr B2 cloud region.

We estimate the total X-ray luminosity to explain the intensity of the scattered X-rays. Assuming that Sgr B2 cloud is a scattering material, optical depth of the cloud is estimated to be 0.2–0.3 from the mass ($\sim 2\times10^6$ M_{solar}) and size ($\sim 10'\times10'$) of Sgr B2 cloud (e.g., /13/). The absorption corrected luminosity of the scattered component in the Sgr B2 region is estimated to be $\sim 5\times10^{35}$ erg s^{-1}.

One candidate of the irradiating X-ray source is galactic X-ray sources located near the Galactic Center. Using the solid angle of the Sgr B2 cloud from the source, we estimated the required luminosity. In this case, the luminosity of larger than 10^{38} erg s^{-1} is about 100 times brighter than the observed luminosity as shown in table 2. Moreover, the value exceeds the Eddington luminosity for a 1 solar mass star. Thus, galactic binary X-ray sources are considered to be difficult to explain the luminosity of the scattered component.

We suspect that the X-ray source is a peculiar object, possibly massive black hole at the Galactic Center. This is another possibility. Since the solid angle of the this region from the Galactic Center is about 0.06 str, the X-ray luminosity of the X-ray source to explain the scattered component is estimated to be larger than 10^{38} erg s^{-1} which is less than the Eddington luminosity of the massive source with $M \sim 10^6$ M_{solar} /1/. From the similarity of the spectrum, our possible interpretation is that our Galaxy has an invisible bright X-ray source at the Galactic Center like a type II Seyfert galaxy.

TABLE 2 X-ray luminosity required to explain the scattered component

Candidate source	Required luminosity (2–10 keV)	Observed luminosity (2–10 keV)
1E1743.1–2843	$\sim 2\times10^{38}$	$\sim 2\times10^{36}$
Hard source	$\sim 6\times10^{38}$	$\sim 8\times10^{35}$
A1742–294	$\sim 1\times10^{39}$	$\sim 3\times10^{36}$
1E1740.7–2942	$\sim 3\times10^{39}$	$\sim 3\times10^{36}$
GX3+1	$\sim 4\times10^{39}$	$\sim 10^{38}$ (Ginga result /16/)
G.C.	$\sim 5\times10^{38}$	$\sim 2\times10^{36}$

SUMMARY

Main results of ASCA observations of the Galactic Center are summarized as follows;

1. Many X-ray sources including a hard X-ray point source near Sgr A* are detected. This hard source is likely to be a quiescent state of the transient source A1742–289 discovered by Ariel-V. The X-ray luminosities of detected sources are of order 10^{35}–10^{36} erg s^{-1}.

2. The largely extended diffuse emission with K-shell transition lines form He-like Si, S, Ar, and Fe was confirmed around the Galactic Center. This extended emission is due to an optically thin hot plasma.

3. The 6.4 keV line emission predominantly from the positive Galactic longitude side of the Galactic plane was found. The brightest regions coincide the Sgr B2 cloud and clouds near the Radio Arc. The origin is likely to be fluorescent emission irradiated by very bright X-ray source.

ACKNOWLEDGEMENT

The authors thanks all the member of the ASCA team. This work was partly supported by the Scientific Research Fund of the Ministry of Education, Science, and Culture under Grant No. 06233202.

S. Yamauchi *et al.*

REFERENCES

1. R. Genzel and C. H. Townes, *Ann. Rev. Astr. Ap.*, <u>25</u>, 377 (1987).

2. N. Kawai, E. E. Fenimore, J. Middleditch, R. G. Cruddace, G. G. Frits, W. A. Snyder, and M. P. Ulmer, *Astrophys. J.*, <u>330</u>, 130 (1988).

3. G. K. Skinner, A. P. Willmore, C. J. Eyles, D. Bertram, M. J. Church, P. K. S. Harper, J. R. H. Herring, J. C. M. Peden, A. M. T. Pollock, T. J. Ponman, and M. P. Watt, *Nature*, <u>330</u>, 544 (1987).

4. M. G. Watson. R. Willingale, J. E. Grindlay, and P. Hertz, *Astrophys. J.*, <u>250</u>, 142 (1981).

5. R. Sunyaev, E. Churazov, M. Gilfanov, M. Pavlinsky, S. Grebenev, G. Babalyan, I. Dekhanov, N. Yamburenko, L. Bouchet, M. Niel, J. -P. Roques, P. Mandrou, A. Goldwurm, B. Cordier, Ph. Laurent, and J. Paul, *Astron. Astrophys.*, <u>247</u>, L29 (1991).

6. K. Koyama, H. Awaki, H. Kunieda, S. Takano, Y. Tawara, S. Yamauchi, I. Hatsukade, and F. Nagase, *Nature*, <u>339</u>, 603 (1989).

7. S. Yamauchi, M. Kawada, K. Koyama, H. Kunieda, Y. Tawara, and I. Hatsukade, *Astrophys. J.*, <u>365</u>, 532 (1990).

8. P. von Ballmoos, R. Diehl, and V. Schöfelder, *Astrophys. J.*, <u>318</u>, 654 (1987).

9. Y. Tanaka, H. Inoue, and S. S. Holt, *Publ. Astron. Soc. Japan*, <u>46</u>, L37 (1994).

10. R. Sunyaev, E. Churazov, M. Gilfanov, M. Pavlinsky, S. Grebenev, G. Babalyan, I. Dekhanov, N. Khavenson, L. Bouchet, P. Mandrou, J. -P. Roques, G. Vedrenne, B. Cordier, A. Goldwurm, F. Lebrun, and J. Paul, *Astrophys. J.*, <u>383</u>, L49 (1991).

11. C. J. Eyles, G. K. Skinner, A. P. Willmore, and F. D. Rosenverg, *Nature*, <u>257</u>, 291 (1975).

12. A. M. Willson, G. F. Carpenter, C. J. Eyles, G. K. Skinner, and A. P. Willmore, *Astrophys. J.*, <u>215</u>, L111 (1977).

13. D. C. Lis and J. E. Carlstrom, *Astrophys. J.*, <u>424</u>, 189 (1994).

14. H. Awaki, this issue.

15. S. Ueno, R. F. Mushotzky, K. Koyama, K. Iwasawa, H. Awaki, and I. Hayashi, *Publ. Astron. Soc. Japan*, <u>46</u>, L71 (1994).

16. K. Asai, Ph. D. thesis, Science University of Tokyo (1994).

Pergamon

PII: S0273-1177(97)00039-2

Adv. Space Res. Vol. 19, No. 1, pp. (1)71–(1)74, 1997
Published by Elsevier Science Ltd on behalf of COSPAR
Printed in Great Britain
0273–1177/97 $17.00 + 0.00

HIGH-ENERGY SPECTRUM OF THE GALACTIC CENTER BLACK HOLE Sgr A

Fulvio Melia

Depatment of Physics and Steward Observatory, University of Arizona, Tucson AZ 85721, U.S.A.

ABSTRACT

The massive blackhole candidate Sgr A* may be accreting from an ambient Galactic center wind at a rate $\approx 10^{22}$ g s^{-1}. Dissipative processes within the large-scale quasi-spherical infall, from $50\, r_g$ to $10^5\, r_g$ (where $r_g = 3 \times 10^{11}$ cm is the Schwarzschild radius for $M = 10^6\, M_\odot$) can account well for the observed radio spectrum and flux (due to cyclotron/synchrotron emission) and its X-ray/Gamma-ray luminosity (apparently due to Bremsstrahlung emission). Small scale instabilities associated with the stagnation region of the flow produce fluctuations in the accretion rate with an amplitude of up to 30% and a time scale of several months to over a year. This may account for the long term variability of the high-energy (and radio) luminosity observed from Sgr A*. Although the average accreted angular momentum is approximately zero, these instabilities also induce fluctuations in the specific angular momentum of the accreted gas that lead to a circularized flow at distances less than about $20\, r_g$. Optically thick emission from this "disk" (roughly the size of Mercury's orbit) is the origin of the IR flux recently detected from the location of Sgr A*, but is not itself a significant source of high-energy emission.
Published by Elsevier Science on behalf of COSPAR

INTRODUCTION

The uniqueness of the radio source Sgr A*, together with its low proper motion ($\lesssim 40$ km s^{-1} /1/) and its location near the dynamical center of the galaxy /2,3/ suggest that it may be a massive point-like object dominating the gravitational potential in the inner $\lesssim 0.5$ pc region /4/. Over the past several years, multi-wavelength observations of the galactic center region have provided a detailed picture of the gas dynamics in the vicinity of this object, pointing to an accretion model as perhaps the most likely physical process producing its spectral properties (see also /5/ for an earlier discussion). In conjuction with related measurements of high outflow velocities associated with infrared sources in Sgr A West /6/, and the interpretation of the H$_2$ line from the inner edge of the circumnuclear ring surrounding Sgr A West as the emission from molecular gas being shocked by a nuclear mass outflow /7/, observations of broad He I, Brα, and Brγ emission lines from the vicinity of IRS 16 /8/, and by Ne$^+$ /9/ and Radio continuum /10/ observations of IRS 7 provide clear evidence for the presence of a very strong wind (with velocity $v_{gw} \approx 500 - 700$ km s^{-1} and a mass-loss rate $\dot{M}_{gw} \approx 3 - 4 \times 10^{-3}\, M_\odot$ yr^{-1}) pervading the inner few parsecs of the galaxy. Most, perhaps all, of Sgr A*'s radiative characteristics may be due to the energy liberated by this wind as it accretes down the deep potential well.

In this paper, we summarize the results of some recent developments in this model, focusing on the key questions that have been (or are in the process of being) addressed, such as (1) what is the physical state of the infalling gas? (2) what is the corresponding emissivity (i.e., the relative contribution to the spectrum due to bremsstrahlung, magnetic bremsstrahlung, etc.)? (3) does the accreting plasma carry specific angular momentum, e.g., to form a disk? and (4) what is the overall flux from this object as a function of mass, given the (fairly well-known) wind parameters associated with the ambient gas?

---

[1] Presidential Young Investigator.

PHYSICAL STATE OF THE GAS AND THE EMITTED SPECTRUM

The accretion is initiated by a bow shock that dissipates most of the directed kinetic energy and heats the gas to a temperature $T_0 = (3\mu_H\, m_H/16k)v_{gw}^2 \approx 7 \times 10^6$ K $(v_{gw}/600$ km s$^{-1})^2$. The accretion rate $\dot{M}$ is roughly proportional to the solid angle subtended at the wind source by a circle of radius $r_a \equiv 2GM/v_{gw}^2$ (the "accretion radius"), within which the directed particle kinetic energy is less than the gravitational potential energy. Thus, $\dot{M} \approx 1 \times 10^{22}$ g s^{-1} $\mathcal{W}\,(M/10^6\,M_\odot)^2$, where $\mathcal{W} \equiv (v_{gw}/600\,\text{km s}^{-1})^{-4}\,(\dot{M}_{gw}/3.5 \times 10^{-3}\,M_\odot\,\text{yr}^{-1})\,(D/0.06\text{ pc})^{-2}$, and D is the deprojected distance between Sgr A* and the (effective) origin of the Galactic-center wind.

The physics of (quasi-)spherical accretion onto blackholes has been the subject of many studies, dating back to the original work by /11/. Since then, much of the groundwork has been developed by, e.g., /12,13,14/, among others. As discussed therein (and /15,16/), the magnetic field within the accreting, highly ionized plasma is intensified by compression and flux conservation, so that in addition to radiating via bremsstrahlung emission, the particles also radiate due to their orbital motions around the magnetic lines of induction. However, the frequency and emissivity of the radiation undergo dramatic changes as the particle energy makes a transition from nonrelativistic to extreme relativisitc values /16/, due primarily to the ever-greater smearing of the successive harmonics, which causes a pronounced overlapping that results in an almost monotonically decreasing spectrum at high temperature, but only a series of discrete lines at low energy.

Reasonable estimates for the density and velocity in the infalling gas may be obtained by assuming conservation of baryon number, which together with the "equipartition" argument /11/ then also give the distribution of magnetic field. Application of the first law of thermodynamics to the particle specific energy in the fluid frame yields the temperature, which completes the specification of the physical state of the gas /15,16/. If the infalling plasma retains (or acquires) some specific angular momentum $l \equiv \lambda\,cr_g$, the inward velocity approaches zero at the "circularization" radius $r_{circ} \equiv 2\lambda^2\,r_g$, where the matter settles down into a Keplerian orbit. The radiative emission from this (small) disk dominates the spectrum at IR and UV wavelengths, whereas the bremsstrahlung and magnetic bremsstrahlung emissivities from the extended quasi-spherical infall are the dominant contributors to the radio, X-ray and possibly γ-ray flux.

Figure 1 shows the calculated run of temperature as a function of radius (in units of r_g) for three values of the angular momentum parameter λ. With these distributions thus determined, it is straightforward to calculate the expected spectrum across 14 decades in frequency (Figure 2). On this plot are also included the available data for comparison, assuming a distance $D = 8.5$ kpc to the galactic center. Below a frequency $\sim 10^{12}$ Hz, the luminosity is entirely dominated by magnetic bremsstrahlung emissivity within the quasi-spherical accretion flow, but this contribution is insignificant at higher photon energy due to the much lower cyclotron/synchrotron opacity in this frequency range. In other words, the medium becomes transparent in the far IR region of the spectrum, allowing us to see the contribution from the disk in the IR, optical, and UV wavebands, though of course the latter two are heavily extinguished in the plane of the galaxy. Other than this contribution from the disk, the spectrum above $\sim 10^{12}$ Hz is dominated by electron-ion and electron-electron bremsstrahlung in the extended infall. It is important to emphasize that the bremsstrahlung and magnetic bremsstrahlung emissivities, and the disk luminosity using the same accretion rate $\dot{M}$, are calculated self-consistently from the same distribution of particle density and temperature. As such, the reasonable agreement with the data across 12 decades in frequency is significant.

DISCUSSION

In the course of a full-sky survey, the EGRET detector on the Compton GRO has detected a ~ 30 MeV - 30 GeV γ-ray excess that appears to originate from within the inner ~ 400 pc of the Galaxy /17/. If the source is at the Galactic center, the inferred γ-ray luminosity density is $\lesssim 8 \times 10^{11}$ ergs s^{-1} Hz^{-1}, which might represent the high-energy extension of the bremsstrahlung shoulder in Figure 2. Very importantly, the X-ray to γ-ray emission appears to be highly variable (on a time

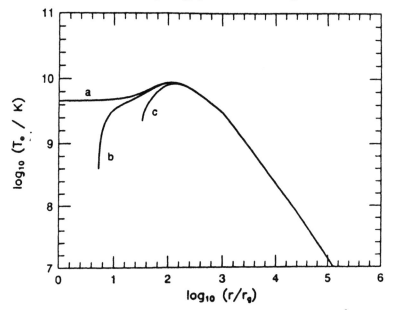

Fig. 1. T_e as a function of radius r for a blackhole mass $M = 2 \times 10^6 \, M_\odot$ and wind parameter $\mathcal{W} = 0.3$, corresponding to a total accretion rate $\dot{M} \approx 1.2 \times 10^{22} \, g \, s^{-1}$ onto Sgr A*. Curve a) is for $\lambda = 0$, b) $\lambda = 2$, and c) $\lambda = 5$.

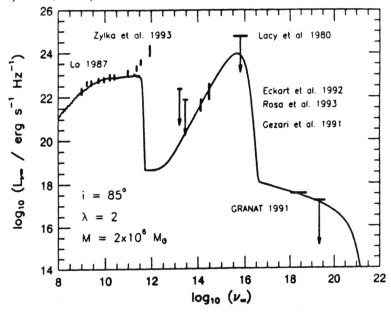

Fig. 2. The predicted luminosity as a function of frequency (at infinity), for $M = 2 \times 10^6 \, M_\odot$ and $\mathcal{W} = 0.3$. λ is the assumed (dimensionless) specific angular momentum of the accreting plasma and i = disk inclination angle.

scale of 6 months, or less), which appears to support the single source hypothesis. In addition, the confrontation between this model and the IR observations suggests that the quasi-spherical infall eventually produces a (small) disk near the blackhole (see the $\sim 10^{14}$ Hz portion of the spectrum in Fig. 2). Motivated in part by how these conditions (i.e., the variability and small disk formation) might be achieved, we have initiated a study incorporating 3D hydrodynamical simulations of the accretion process to better understand the structure of Sgr A*'s "halo" /18/. In Figure 3 we show some preliminary results of these calculations, which indicate that most of the angular momentum l is cancelled in the post-accretion-shock region, though local fluctuations in the accretion rate produce transient excesses in l with a characteristic time scale $\lesssim 10$ years.

F. Melia

Interestingly, the average magnitude of these fluctuations appears to correspond to a $\lambda \approx 3$, close to the value inferred from the spectral fit in Figure 2. On the basis of this model, we should therefore expect significant variations in the K, H, X-ray and γ-ray flux from Sgr A* over a period of several years or less, though with possible frequency-dependent delays arising from differences in the location of the emission regions. Future studies that incorporate both detailed hydrodynamical simulations and new constraints from multi-frequency correlated observations should allow us to map this environment near one of the most interesting objects in the Galaxy.

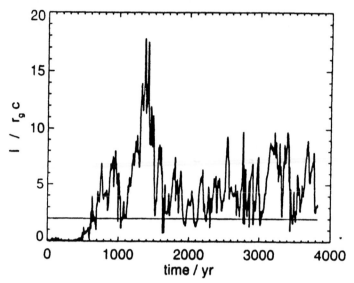

Fig. 3. *The absolute specific angular momentum l accreted by the black hole as a function of time, for a uniform (Mach 30) gas flowing past an object of mass $M = 1 \times 10^6 \, M_\odot$. The thin horizontal line (at $l = 2r_g c$, i.e., $\lambda = 2$) represents the value of l required to account for the IR luminosity shown in Fig. 2. Thus, the small (transient) accretion disk surrounding Sgr A* may simply be due to the fluctuations in specific angular momentum accreted by the infalling gas.*

This work was supported in part by NSF grant PHY 88-57218 and NASA grant NAGW-2518.

REFERENCES

1. Backer, D.B. & Sramek, R.A. 1987, in AIP Proc. 155, ed. D.C.
2. Rieke, G.H. & Rieke, M.J. 1988, Ap.J.Lett. 330, L33.
3. Lacy, J.H., Achtermann, J.M. & Serabyn, E. 1993, Ap.J.Lett., in press.
4. Lynden-Bell, D. & Rees, M. 1971, *M.N.R.A.S.*, 152, 461.
5. Ozernoy, L.M. 1989, in The Center of the Galaxy, ed. M. Morris (Kluwer: Dordrecht), p. 555.
6. Krabbe,A., Genzel, R., Drapatz, S., and Rotaciuc, V. 1991, Ap.J.Lett., in press.
7. Gatley, I., Jones, T.J., Hyland, A.R., Wade, R., Geballe, T.R, & Krisciumas, K. 1986, M.N.R.A.S., 222, 562.
8. Hall, D.N.B., Kleinmann, S.G. and Scoville, N.Z. 1982, Ap.J.Lett., 260, L63.
9. Serabyn, E., Lacy, J.H., & Actermann, J.M. 1991, Ap.J., 378, 557.
10. Yusef-Zadeh, F. & Melia, F. 1991, Ap.J.Lett., 385, L41.
11. Shvartsman, V.F. 1971, Soviet Astron.-AJ, 15, 377.
12. Mészarós, P. 1975, Astro. Ap., 44, 59.
13. Schmid-Burgk, J. 1978, Ap. Space. Sci., 56, 191.
14. Ipser, J.R. & Price, R.H. 1983, Ap.J., 267, 371.
15. Melia, F. 1992, Ap.J.Letters, 387, L25.
16. Melia, F. 1994, Ap.J., 426, 577.
17. Mattox, J.R., et al. 1993, BAAS, 24, 1296.
18. Ruffert, M. & Melia, F. 1994, Astron. Astrop. in press.

Pergamon

Adv. Space Res. Vol. 19, No. 1, pp. (1)75–(1)84, 1997
© 1997 COSPAR
Printed in Great Britain. All rights reserved
0273–1177/97 $17.00 + 0.00

PII: S0273-1177(97)00040-9

RAPID X-RAY VARIABILITY IN GALACTIC BLACK HOLE CANDIDATES

M. van der Klis

Astronomical Institute "Anton Pannekoek" and Center for High-Energy Astrophysics, Kruislaan 403, 1098 SJ Amsterdam, The Netherlands

ABSTRACT

The rapid X-ray variability of galactic accreting black-hole candidates is compared to that of accreting neutron stars with low magnetic fields. The power spectra of these objects can be described in terms of a small number of simple power spectral shapes: power law noise, band limited noise, and quasi-periodic oscillations (QPO). In a given source, the properties of these power spectral components seem to depend, to first order, only on mass flux. Similarities in the power spectral properties between source types strongly suggest that similar physical mechanisms underlie power spectral components seen in black-hole candidates and in neutron stars with various magnetic-field strengths. Two rapid ($\gtrsim 1$ Hz) QPO and two band limited noise components appear to occur across all types of X-ray binaries; the situation with respect to the ubiquitous power law components, as well as with respect to slow ($\lesssim 1$ Hz) QPO is as yet unclear. One QPO and one band limited noise component appear to be magnetospheric as they are not seen in black-hole candidates and atoll sources (which are inferred to be neutron stars with a very low magnetic field). Another QPO and band limited noise component may be related to the presence of an inner, radiation pressure dominated accretion disk, as they do not occur in X-ray pulsars. It is discussed to what extent the relatively small *quantitative* differences between the rapid X-ray variability properties of neutron stars and black-hole candidates can be used to identify black holes, and whether there exist any *qualitative* differences (*i.e.*, black hole signatures) in the rapid X-ray variability.

1. INTRODUCTION

The historical record of black hole "signatures" that were subsequently associated with neutron star systems (bimodal spectral states, rapid variability, see, *e.g.*, Tanaka 1989) testifies to the fact that the accretion process onto neutron stars and black holes has much in common. This is to be expected, as the characteristics that make accreting compact objects unique, namely, the very high, and thinly shielded, energy density (the origin of the high energy emission), and the short dynamical and radiative time scales (underlying the rapid [millisecond] variability), are common to neutron stars and black holes. To first order one would expect the accretion flow onto these two types of compact object to be very similar; only in the very innermost region (a few gravitational radii from the center) would one expect the differences between neutron stars and black holes (such as the presence or absence of a material surface or relativistic frame dragging) to become noticeable. It should be kept in mind, however, that this innermost region is the region where most of the gravitational energy is released, and that at least in principle the effects of what happens in the middle could propagate outward and affect the accretion flow at larger distances.

It might be expected, then, that the phenomena displayed by accreting neutron stars and black holes are similar, but that the parameters needed in describing these phenomena differ: phenomenologically the differences between the two groups might be more quantitative than qualitative. With this in mind, in the following I review the X-ray observations of the black-hole candidates and compare them to the low magnetic-field neutron stars, with particular emphasis on the observations of their rapid X-ray variability. Indeed, similarities have emerged that indicate that a unified description of some of the accretion phenomena may be possible. If a phenomenon is seen in both neutron star and black-hole candidate systems this tells us quite a bit about the phenomenon, as then it can not be due to any property that is unique to either neutron stars or black holes, such as the presence or absence of a surface, or the presence or absence of a strong non-aligned magnetic field. In addition, some source characteristics are emerging from the studies of the similarities of neutron stars and black holes, that may indeed be unique to black holes.

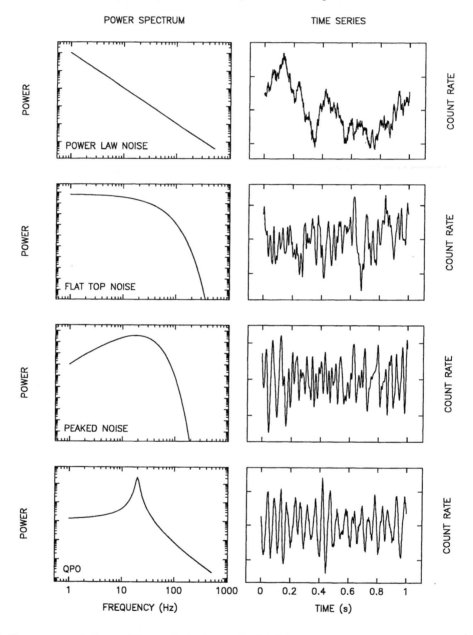

Fig. 1. Power-spectral shapes. The synthetic time series (right) were calculated by inversely transforming a synthetic Fourier transform with amplitudes given by the square root of the assumed power spectrum (left) and random phases. No counting noise was included.

In the process of accretion onto a compact object, the X-ray spectrum and the rapid X-ray variability originate in the same physical region (near the compact object), so that these X-ray properties are expected to be coupled. The hypothesis that the mass flux $\dot{M}$ towards the compact object governs both the X-ray spectrum and the power spectrum, so that when $\dot{M}$ varies these observables will show correlated variations works well in explaining the data. A very recent development is that maybe we are beginning to understand some of the effects of binary inclination in the non-dipping LMXB; I'll briefly discuss how the results from these sources might be applied to black-hole candidates in Section 4. For more extensive reviews I refer to van der Klis (1994b, d).

In the following I shall use the expression "black-hole candidate" not only for compact objects that (from orbit measurements) have a mass too high for a neutron star (the black-hole candidates proper), but also for those that phenomenologically appear to fall in the same group as these sources (*e.g.*, Tanaka 1992, Grebenev et al. 1993), the "candidate black-hole candidates" or black-hole candidates "by association".

The power spectra of accreting compact objects can be described in terms of a small number of simple shapes (Fig. 1). *Power law noise* has a power distribution $\propto \nu^{-\alpha}$, *band limited noise* one that steepens towards high ν and flattens towards low ν. Band limited noise that has a maximum at $\nu > 0$ is called *peaked*; if the maximum is at $\nu = 0$ the component is called *flat-topped*. The same power spectral component can be at one time flat-topped and at another time peaked. *Quasi-periodic oscillations* (QPO) are a type of peaked noise. Usually, the term QPO is reserved for relatively narrow peaks.

2. BLACK HOLE STATES

Three source states are distinguished in black-hole candidates (Tananbaum et al. 1972, Oda et al. 1976, Miyamoto et al. 1991). In the *low state* (LS) the X-ray spectrum is a flat power law with photon spectral index 1.5–2. In the *high state* (HS) the 1–10 keV flux is much higher due to a soft component; the power law is sometimes "sticking out" from under the soft component at higher energies. In the *very high state* (VHS) the X-ray spectrum is similar to that in the high state (at higher 1–10 keV flux), with perhaps an additional hard power law component. The VHS is mainly distinguished from the HS by the properties of its rapid X-ray variability.

Fig. 2 summarizes the power spectra in the three states. The LS power spectrum shows strong (30–50% amplitude) band-limited noise with ν_{cut} between 0.03 and 0.3 Hz. This LS noise is usually flat-topped, but sometimes peaked (Vikhlinin et al. 1994). The level of the flat top and, in anti-correlation with this, the cut-off frequency ν_{cut} sometimes vary, whereas the power spectrum above ν_{cut} remains approximately unchanged (Belloni and Hasinger 1990, Miyamoto et al. 1992a). In the HS power law noise with $\alpha \sim 1$ and an amplitude of a few % is present. Sometimes LS noise is present in the hard X-ray spectral component seen in the HS, and this is taken as evidence that the LS noise is a property of the hard power-law X-ray spectral component. Slow QPO with frequencies similar to the LS noise cut-off frequencies ($\sim$0.08–0.8 Hz; Motch et al. 1983, Ebisawa et al. 1989, Grebenev et al. 1991) and possibly related to peaked LS noise sometimes occur in LS and HS. The rare VHS shows 3–10 Hz QPO and rapidly variable broad-band noise. The QPO show second harmonics and possible subharmonics. The noise in the VHS alternates in shape, sometimes within 1 s, between band-limited ($\nu_{cut} \sim$1–10 Hz), and power law shaped ($\alpha \sim$1). CD/HID branches occur in the VHS, and the power spectral parameters seem to depend on position in the branches, but these branch structures are not very similar from one epoch to the next ("messy" branches). The LS and VHS band limited noise cut-off frequency and amplitude fit a single relation (van der Klis 1994c), suggesting that they part of a single phenomenon. The VHS power law noise is similar to that in the HS.

The transient black-hole candidate GS 1124−68 (Nova Mus '91) in its decay went through all three states (Miyamoto et al. 1992b, Kitamoto et al. 1994), strongly suggesting that the states directly follow $\dot{M}$, an assumption that works very well in the bright low magnetic-field neutron-star systems. Recent evidence indicates that some black hole transients, even when they are very luminous, remain in the "low" state; see Sunyaev, these proceedings.

BLACK–HOLE–CANDIDATE POWER SPECTRA

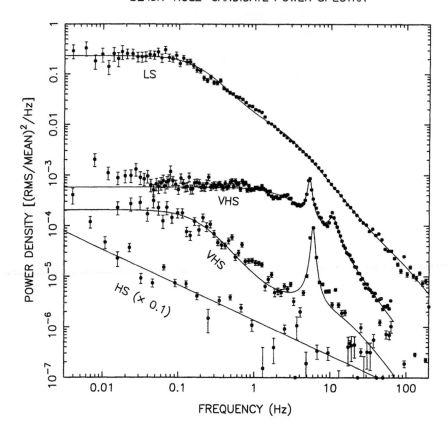

Fig. 2. Power spectra from Ginga data of black-hole candidates in the low (LS; Cyg X-1), high and very high (HS and VHS; GS 1124–68) states.

3. SIMILARITIES BETWEEN BLACK HOLE CANDIDATES AND LOW MAGNETIC FIELD NEUTRON STARS

There is a number of striking similarities between black-hole candidate and neutron star phenomenology (see van der Klis 1994a, d). To describe these similarities it is necessary to briefly review the phenomenology of the accreting low magnetic-field neutron stars, the Z and atoll sources.

Z and atoll sources (Hasinger and van der Klis 1989, hererafter HK89, see also van der Klis 1989, 1994b and references therein) show rather subtle X-ray spectral changes. *Colour-colour diagrams* (CDs) and *hardness-intensity diagrams* (HIDs), plots of X-ray hardness ratios *vs.* each other or *vs.* count rate are used to describe the X-ray spectral variations. The sources produce a characteristic track in the CD/HIDs, and source position in the track is used as an indication for $\dot{M}$.

Z sources produce tracks in the CD/HIDs that are Z-shaped. $\dot{M}$ is inferred to increase following the Z track from upper left to lower right. Z source power spectra show three broad noise components, *very low frequency noise* (VLFN), *low frequency noise* (LFN) and *high frequency noise* (HFN) and two QPO components, *horizontal branch oscillations* (HBO) and *normal and flaring branch oscillations* (N/FBO). VLFN is 1–6% amplitude power law noise that gets stronger with $\dot{M}$. HBO and LFN are a QPO and a band limited noise component that appear an disappear together, and are likely physically related. They are strongest at low $\dot{M}$ and disappear at high $\dot{M}$. HBO frequency (13–55 Hz) and LFN cut-off frequency (2–20 Hz) increase with $\dot{M}$. LFN can be flat topped or peaked, depending on the source. N/FBO have a preferred frequency near 6 Hz, increasing to ~20 Hz when $\dot{M}$ increases.

The most succesful HBO model is the *magnetospheric beat frequency model* (Alpar and Shaham 1985, Lamb et al. 1985), which requires Z sources to have a magnetosphere. In most models for the N/FBO, *radiation pressure* plays the key role (van der Klis et al. 1987, Hasinger 1987, Lamb 1989, Fortner et al. 1989, Miller and Lamb 1992, Alpar et al. 1992). Z sources have near-Eddington luminosities, and the NBO frequency is roughly similar in each Z source, in accordance with the idea that the frequency is determined by the Eddington critical luminosity L_{Edd}. Lamb (1991) proposed a comprehensive model for the QPO and X-ray spectral properties of Z sources that uses the above ingredients.

Atoll sources (HK89, van der Klis 1994b) show one curved branch in the CD, often fragmented due to observational effects. $\dot{M}$ increases from left to right along the branch. Their power spectra show two broad noise components called *very-low-frequency noise* (VLFN) and *high-frequency noise* (HFN). Atoll source VLFN is power law noise similar to that in Z sources. Atoll source HFN has a cut-off frequency of 0.3–20 Hz and depends strongly on $\dot{M}$. At low $\dot{M}$ it is strong (up to 22%); when $\dot{M}$ increases this decreases to <2% (HK89) while the cut-off frequency increases (Yoshida et al. 1993, Prins et al. 1994). Atoll HFN is sometimes flat-topped and sometimes peaked.

HK89 proposed that the neutron stars in atoll sources have lower magnetic field strengths than Z sources, and are constrained to lower mass fluxes $\dot{M}$. The lower field explains why the (magnetospheric) HBO are not seen in atoll sources, and the lower $\dot{M}$ why the same is true for the (near-Eddington) N/FBO. The implied relation between $\dot{M}$ and magnetic field strength may have an evolutionary origin (van der Klis 1991). Predictions are that an atoll source that becomes bright will show Z source high-$\dot{M}$ properties (N/FBO and appropriate spectral branches), but never HBO, and that a Z source that becomes faint will show millisecond pulsations.

The properties of Cir X-1 fit the first prediction. This source is a low magnetic-field neutron star (it shows type 1 X-ray bursts; Tennant et al. 1986a, b) with a complex phenomenology that most likely originates in the large variations in mass transfer that the system undergoes as a function of its 17-d period. Sometimes (at intermediate brightness levels and away from periastron) its power-spectrum and CD behaviour are very similar to those of an atoll source on the banana branch (Oosterbroek et al. 1994). When the source becomes very bright, at periastron, it sometimes shows 6–20 Hz QPO and spectral branches that are reminiscent of Z source N/FBO behaviour (Tennant 1987; Makino et al. 1992; Oosterbroek et al. 1994). The source is apparently an example of an atoll source that can reach $\dot{M}_{Edd}$ (van der Klis 1991; Oosterbroek et al. 1994). As will be discussed below Cir X-1 also shares some characteristics with black-hole candidates.

I now turn to the comparison of the above-described neutron star properties with those of the black-hole candidate systems described in the previous section.

The black-hole candidate LS is very similar to the atoll source low $\dot{M}$ ("island") state. Both states occur at the lowest 1–10 keV count rates and inferred $\dot{M}$ levels. Both are dominated by strong (several 10%) band limited noise (LS noise and atoll HFN) which is sometimes flat-topped and sometimes slightly peaked. When an atoll source becomes really faint, the power spectra are nearly indistinguishable (Fig. 3) from a black-hole candidate in the low state and the 1–20 keV (Langmeier et al. 1987, Yoshida et al. 1993) and 13–80 keV (Van Paradijs and van der Klis 1994) X-ray spectra become hard, just as in black-hole candidates in the LS. Even the inverse correlation between cut-off frequency and flat-top level, characteristic for black-hole candidates in the LS, was seen in an atoll source, 4U 1608−52, at low $\dot{M}$ (Yoshida et al. 1993). Z source LFN fits in with black-hole candidate LS noise and atoll HFN: it is also stronger at lower $\dot{M}$, disappears at higher $\dot{M}$, can be peaked and flat-topped, and has a higher ν_{cut} at higher $\dot{M}$. The absence of a similar band-limited noise component in pulsars, and also the beat-frequency model as applied to Z sources, suggest that such noise arises through inhomogeneities in the inner, radiation pressure dominated part of the disk, which in pulsars is disrupted by magnetic stresses.

The black hole candidate VHS has strong similarities to the Z source high $\dot{M}$ ("normal/flaring branch") state. Both occur at the highest inferred $\dot{M}$ levels, and both show QPO, with similar frequencies (6–20 Hz in the neutron star systems, 3–10 Hz in the black-hole candidates), that depend on the position of the source in branched tracks in the HID/CDs. Clearly different is the harmonic content of the QPO (black-hole candidate VHS QPO show strong harmonics, Z source N/FBO do not) and the character of the HID/CD branches (much "messier" in black-hole candidates). Another difference is that Z sources do not show the fast changes in broad band noise shape seen in black-hole candidates.

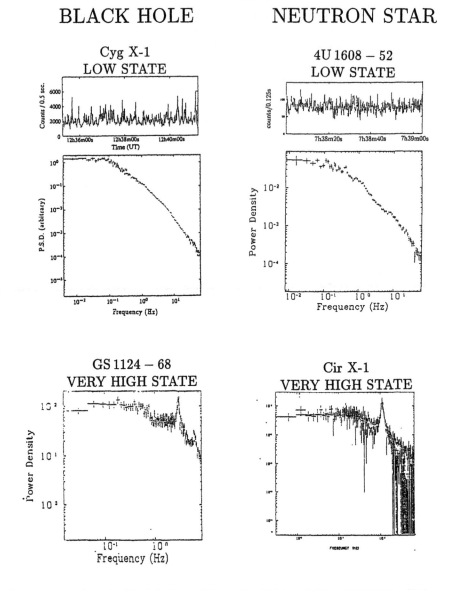

Fig. 3. Power spectra from the black-hole candidates Cyg X-1 (top left) and GS 1124–68 (bottom left) in the low state and the very high state, respectively, and from the low magnetic-field neutron stars 4U 1608–52 (top right) and Cir X-1 (bottom right) in the atoll island state and a very high X-ray brightness state respectively, illustrating the similarity between neutron star and black-hole candidate low and very high states. Compiled from Inoue (1992), Takizawa et al. (1994) and Makino et al. (1991).

The properties of Cir X-1 provide a further link between neutron stars and black holes. In some of its high states (Tennant 1987, Makino et al. 1992, Oosterbroek et al. 1994), this source shows a mix of characteristics of Z sources and black-hole candidates in high $\dot{M}$ states (see Fig. 3). It shows QPO with frequencies between 6 and 20 Hz and no second harmonics (both Z source characteristics) in combination with messy branches in the CD/HID and fast changes in the shape of the broad band noise (black-hole candidate characteristics). In the present interpretation, the reason that Cir X-1 sometimes resembles a black hole in its rapid variability characteristics, as was noted by Toor (1977) and Samimi et al. (1979), while its X-ray bursts show it to be a neutron star, is that it is the only neutron star that we know that has a magnetic field as low as in atoll sources that sometimes accretes at near- or super-Eddington rates. Cir X-1 is therefore a key object as it can help to distinguish between phenomena that are characteristic for accretion onto any compact object that has no appreciable magnetic field, and phenomena that are truly characteristic for accretion onto a black hole. Following this line of reasoning, one concludes that a high harmonic content of the high $\dot{M}$ QPO may be a black hole signature, whereas variable broad band noise and messy branches are not.

On the basis of this array of similarities it can be concluded that the phenomenology of the black-hole candidates and low magnetic-field neutron stars may be described in terms of three $\dot{M}$-driven states that are common to accreting low magnetic-field neutron stars and accreting black holes (van der Klis 1994a). Fig. 4 presents a line-up of the three common states of black-hole candidates and low magnetic-field neutron stars.

Phenomena / Guess at B	Black hole (0 Gauss)	Atoll source ($\lesssim 10^{9-10}$ Gauss)	Z source ($\sim 10^{9-10}$ Gauss)	Guess at $\dot{M}$
High $\dot{M}$ QPO	VHS	Cir X-1 (high state)	FB	$\gtrsim 1\dot{M}_E$
Weak power-law noise			NB	$\sim 0.9\dot{M}_E$
		Banana		
Weak power-law noise + band-limited noise	HS		HB	$\sim 0.5\dot{M}_E$
Strong band-limited noise	LS	Island	(ms X-ray pulsars?)	$\sim 0.01\dot{M}_E$
		(most bursters and dippers)		

VHS: very high state; HS: high state; LS: low state; FB: flaring branch; NB: normal branch; HB: horizontal branch.

Fig. 4. *Proposed classification scheme for X-ray binary source states. There a three states that are common to neutron stars and black holes; in a given source the mass transfer rate $\dot{M}$ towards the compact object determines the state. The power spectral shapes that are characteristic of each state are indicated in the leftmost column. The correspondence between the source states of each source type is indicated. Magnetic field strengths and mass fluxes are rough indications only. In particular, other source parameters, in particular inclination, are expected to affect the $\dot{M}/\dot{M}_E$ levels at which state transitions occur.*

It was proposed recently that the power law noise components (VLFN) seen in accreting neutron stars might be due to unsteady nuclear burning on the neutron star surface (Bildsten 1993). A prediction of the model is that no VLFN similar to that in neutron stars should be seen in black holes.

4. DIFFERENCES BETWEEN BLACK HOLE SYSTEMS: INCLINATION EFFECTS?

The situation with respect to the states of black-hole candidates, and their relation to $\dot{M}$ is not yet entirely clear. In particular, there are cases, such as GX 339−4, where the observable energy flux in the 1–200 keV band is higher in the low state than in the high state (see Fig. 3 in Grebenev et al. 1993) and of sources, such as GS 2023+338, that despite an apparently very high X-ray luminosity display clear low-state characteristics (Kitamoto et al. 1989; Inoue 1989; 1991; Terada et al. 1992; as mentioned above, there may be a class of sources that does not show a soft component at all, see Sunyaev, these proceedings).

HARD

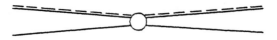

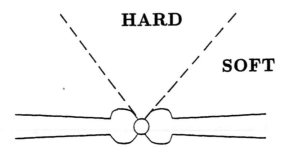

HARD

SOFT

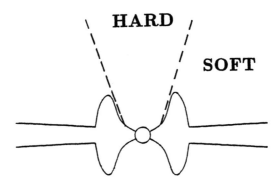

HARD

SOFT

Fig. 5. A geometrically thick inner disk could obscure a central, hot and rapidly variable emitting region for observers at high inclination. $\dot{M}$ increases from top to bottom. The regions into which the hard and soft components are emitted overlap, making it possible to observe both components at the same time.

There are various possible explanations for these discrepancies between observed luminosity and inferred accretion rate. One possibility is that a fourth state exists in black-hole candidates, at higher $\dot{M}$ levels than the VHS, which like the LS is characterized by strong band limited noise and a hard X-ray spectrum and which has erroneously been lumped together with the "real" low state (Nowak 1994). However, there is as yet no evidence for any phenomenological differences between these two hard and strongly variable states, as would be expected in such a model. Other explanations would involve redirecting part of the gravitational energy released in the high state into some channel where it is not observed. Again, there are several possibilities as to where the energy might be redirected (mechanical energy of jets, XUV photons; see Grebenev, these proceedings, conceivably even into the hole); here I shall discuss a possibility that is suggested by a study of the properties of neutron star systems (see also van der Klis 1994a).

A detailed examination of the properties of Z sources, in particular in their flaring branches, has led Kuulkers and van der Klis (1994) to propose that obscuration by a geometrically thick inner accretion disk plays a role in Z source phenomenology. The disk swells when $\dot{M}$ increases, and at some $\dot{M}$ the torus-like inner disk begins to obscure the central emitting regions. In Z sources this produces a distortion of the Z pattern. For higher inclinations i, the obscuration effects set in at lower $\dot{M}$.

A similar model could explain some of the phenomenology of black-hole candidates. The reason for the disappearance of the hard LS X-ray spectral component in the HS may be obscuration of a central, hot and rapidly variable region by matter in a puffed-up inner accretion disk. For a pole-on viewing geometry no obscuration would occur and the system would be completely dominated by the hard, rapidly variable X-ray spectral component at all $\dot{M}$ levels; this might explain the behaviour of GS 2023+338 (van der Klis 1994a) and similar sources that remain always in the "low" state (really, the hard state). Such an inner disk geometry could also cause the soft, much less variable, X-ray spectral component to be more concentrated towards the equatorial plane, thus preventing high i sources from showing an appreciable soft component. The increasing concentration of the hard X-rays towards the (rotation) polar axes with increasing $\dot{M}$ would in this scenario explain why the apparent 1–200 keV luminosity in the LS seems to be sometimes higher than in its HS and VHS: most of the energy would be leaving the system in the HS and VHS along the polar axis and not be seen by us (see Fig. 5).

In the low magnetic-field neutron stars the X-ray flux is an unreliable indicator of $\dot{M}$; the same might turn out to be the case in an even stronger sense in the black-hole candidates. Note, that the mass flux $\dot{M}$ that by hypothesis determines the state is the mass flux *towards* the compact object, just as is the case in the Z sources; at near- and super-Eddington rates, not all of this matter may actually be accreted; jets might for example be formed when $\dot{M}$ becomes high enough.

5. CONCLUSIONS

A picture may be emerging where the millisecond fluctuations in black holes, and in neutron stars with high, low and very low magnetic fields can be understood in common terms. Just two structures determine the basic physics of the accretion process, namely the magnetosphere and the inner (radiation pressure dominated) disk. Z sources have the most complex phenomenology, showing HBO and N/FBO, as well as LFN and HFN, because their magnetic field is weak enough to allow the presence of a mostly undisturbed inner accretion disk (like in black holes and atoll sources) and strong enough to allow the presence of a small magnetosphere (like in pulsars). Black holes and atoll sources (including Cir X-1) have an inner disk and no appreciable magnetosphere and therefore only show an N/FBO-like and an LFN-like component. Inner disk structure causes anisotropic emission and thereby inclination effects are introduced in the phenomenology.

Similar phenomena in black hole candidates and neutron stars that could have similar underlying physical mechanisms include the low state "shot" noise, similar in character to the atoll source "high frequency" and the Z source "low frequency" noise, which could be caused by accretion of inhomogeneities in the inner disk, the very high state 3–10 Hz QPO, similar to Z source (and Cir X-1) 6–20 Hz QPO, the hardening of the X-ray spectra at low $\dot{M}$, and the existence of three source states. Black hole signatures may well exist, but it is important to be very precise and quantitative when claiming one: just the presence of hard (photon index <1.5 in the 30–100 keV range) tails, or of strong (>25% rms with a cut-off <0.05 Hz) band limited noise, is probably *not* a black hole signature, the presence of such tails or such a noise component at a luminosity $>10^{37}$ erg/s may be. The presence of strong harmonics (>0.5 times the fundamental) in 3–10 Hz QPO seen at high ($\sim L_{Edd}$) luminosity is another possible black hole signature.

This work was supported in part by the Netherlands Organization for Scientific Research (NWO) under grant PGS 78-277. I gratefully acknowledge a travel grant of the Leids Kerkhoven-Bosscha Fonds.

REFERENCES

Alpar, M.A., Shaham, J., 1985, Nat 316, 239.
Alpar, M.A., et al., 1992, A&A 257, 627.
Belloni, T., Hasinger, G., 1990, A&A 227, L33.

Bildsten, L., 1993, ApJ , 418,L21.

Ebisawa, K., Mitsuda, K., Inoue, H., 1989, PASJ 41, 519.

Fortner, B., Lamb, F.K., Miller, G.S., 1989, Nat 342, 775.

Grebenev, S.A., et al., 1991, Sov. Astron. Lett. 17(6), 413.

Grebenev, S., et al., 1993, A&AS 97, 281.

Hasinger, G., 1987, A&A 186, 153.

Hasinger, G., van der Klis, M., 1989, A&A 225, 79. [HK89]

Inoue, H. 1989, in *Two Topics in X-ray Astronomy*, Proc. 23d ESLAB Symp., J. Hunt & B. Battrick (eds.), ESA SP-296, p. 783.

Inoue, H. 1991, paper presented at the Texas/ ESO-CERN Symp., Brighton, UK, December 1990, ISAS RN 469.

Inoue, H., 1992, ISAS RN 518, in *Accretion disks in compact Stellar Systems*, J.C. Wheeler (ed.).

Kitamoto, S., et al. 1989, Nat 342, 518.

Kitamoto, S., et al., 1994, in prep.

Kuulkers, E., van der Klis, M., 1994, A&A , submitted.

Lamb, F.K., 1989, Proc. 23rd ESLAB Symposium, ESA SP-296, 215.

Lamb, F.K., 1991, NATO Advanced Study Institute 344, 445.

Lamb, F.K., et al., 1985, Nat 317, 681.

Langmeier, A., Sztajno, M., Hasinger, G., Trümper, J., 1987, ApJ 323, 288.

Makino, Y., Kitamoto, S., Miyamoto, S., 1992, in: Frontiers of X-ray astronomy, Tanaka and Koyama (eds.), Universal Academy Press, Tokyo, p. 167.

Miller, G.S., Lamb, F.K., 1992, ApJ 388, 541.

Miyamoto, S., Kimura, K., Kitamoto, S., Dotani, T., Ebisawa, K., 1991, ApJ , 383,784.

Miyamoto, S., et al., 1992a, ApJ 391, L21.

Miyamoto, S., et al., 1992b, Proc. Ginga Memorial Symp., Makino and Nagase (eds.), ISAS, Tokyo, p. 37.

Motch, C., et al., 1983, A&A 119, 171.

Nowak, M.A., 1994, AIP Conf. Proc. 308, 547.

Oda, M., Doi, K., Ogawara, Y., Takagishi, K., Wada, M., 1976, *Ap. Space Sc.* 42, 223.

Oosterbroek, T., et al., 1994, A&A , in press.

Prins, S., et al., 1994, A&A , in prep.

Samimi, J., et al., 1979, Nat 278, 434.

Takizawa, M., et al., 1994, in prep.

Tanaka, Y., 1989, *Proc. 23d ESLAB Symp. on Two Topics in X-ray Astronomy*, Bologna, Italy, 13–20 September, 1989 (ESA SP-296), p. 3.

Tanaka, Y., 1992, Proc. ISAS Symp. on Ap., Tokyo, February 4-5, 1992, p. 19.

Tananbaum, H., Gursky, H., Kellogg, E., Giacconi, R., Jones, C., 1972, ApJ 177, L5.

Tennant, A.F., 1987, MNRAS 226, 971.

Tennant, A.F., Fabian, A.C., Shafer, R.A., 1986a, MNRAS 219, 871.

Tennant, A.F., Fabian, A.C., Shafer, R.A., 1986b, MNRAS 221, 27p.

Terada, K., et al. 1992, in *Frontiers of X-ray astronomy*, Proc. 23d Yamada meeting, Y. Tanaka & K. Koyama (eds.), Universal Academy Press, Tokyo, p. 323.

Toor, A., 1977, ApJ 215, L57.

Van der Klis, M., 1989, ARA&A 27, 517.

Van der Klis, M., 1991, NATO Advanced Study Institute 344, 319.

Van der Klis, M., 1994a, ApJS , 92,511.

Van der Klis, M., 1994b, AIAP preprint 1993-009; in: X-Ray Binaries, Lewin, van Paradijs and van den Heuvel (eds.), Cambridge University Press, in press.

Van der Klis, M., 1994c, A&A 281, L17.

Van der Klis, M., 1994d, in: Proceedings of NATO ASI, Kemer, Turkey, 1993, in press.

Van der Klis, M., et al., 1987, ApJ 316, 411.

Van Paradijs, J., van der Klis, M., 1994, A&A 281, L17.

Vikhlinin, A., et al., 1994, preprint.

Yoshida, K., et al., 1993, PASJ 45, 605.

 Pergamon

Adv. Space Res. Vol. 19, No. 1, pp. (1)85–(1)94, 1997
© 1997. Published by Elsevier Science Ltd on behalf of COSPAR
Printed in Great Britain. All rights reserved
0273–1177/97 $17.00 + 0.00

PII: S0273-1177(97)00041-0

ON THE OBSERVATIONAL EVIDENCE FOR ACCRETING BLACK HOLES IN QUASARS

Fabrizio Fiore*,** and Martin Elvis**

* *Osservatorio Astronomico di Roma*
** *Harvard-Smithsonian Center for Astrophysics*

ABSTRACT

We review (mainly high energy) observations which could lend support to (or dismiss) the hypothesis of an accreting central black hole in quasars. Direct imaging is not powerful (1). We focus on three main topics: 2) variability as a constraint of the quasar compactness; 3) X–ray continuum; 4) X–ray spectral features, expected from reprocessing of the X–ray radiation from matter near the X–ray source. We argue that the above observations provide a weaker evidence than once thought for a black hole as the engine of quasars. New tests will come from (5) the study of the evolution of the quasar Spectral Energy Distribution (SED) in the framework of models of quasar physical evolution. We present some new results, obtained comparing ROSAT X–ray observations of z=2–3 quasars with previous monitoring of low–z quasars, which represent a first steep in this direction.

©1997. Published by Elsevier Science Ltd on behalf of COSPAR. All rights reserved

1. DO WE HAVE DIRECT PROOFS FOR A BLACK HOLE IN QUASARS?

The longstanding 'best buy' model for the underlying power source in quasars [1] is accretion onto a massive black hole. There are four traditional main areas of observational support for the black hole theory:
(1) IMAGING: VLBI sizes; stellar cusps.
(2) VARIABILITY: X–ray compactness; microlensing.
(3) GALACTIC ANALOGS: similarity with Galactic black hole candidates (BHC).
(4) DYNAMICS: Fe-K line profiles; H–α line profiles; HST stellar velocity dispersions (e.g. M87).
To which we may now add:
(5) EVOLUTION: a new family of tests using characteristic break luminosities and redshifts.
We are not going to give a complete review of all above observational efforts, rather we will focus on a few topics, most of which related with high energy observations, organized in order of decreasing 'strength'. We think that, in order to understand how constraining the observations truly are on the black hole model, it is instructive to compare it to an alternative. The Starburst model of quasars /1/, where the central engine is not a compact object but the core of an elliptical galaxy, is the most developed of competing quasar models. So we use this as our 'sparring partner' throughout.

2. VARIABILITY

The strongest tests and constraints on quasar compactness come from observations of microlensing in the optical band and fast variations in the X–ray band. We discuss these next. But first we must dispense with the false clues introduced by relativistic beaming.

<u>Beamed Sources</u>

---

[1] we shall use 'quasars' to refer to all types of activity including all active galactic nuclei, Seyferts, and both radio–loud and radio–quiet objects. We will distinguish only BL Lacs as a separate class in this papers.

Originally, the large amplitude and rapid variations found in BL Lacs and optical violently variable quasars were taken as strong arguments for a very compact source emission region and against any 'Christmas tree' models involving multiple loosely related sources. This argument is now seen to be applicable only to the Blazar class, since most other quasars show much less violent variations. In the case of the Blazars we are confident, from VLBI observations of super–luminal expansion, that the variability is an optical illusion caused by jets of material streaming toward us at relativistic velocities. While dramatic these do not directly constrain the nature of the central source or sources. The observation of highly collimated jets on scales down to ≈ 0.1 pc suggests the existence of preferential directions in the propagation of radio–emitting plasma, and is usually associated with the spin axis of the black hole. However, the size of the VLBI jet 'core' is the order of 0.1 parsec or more in most strong radio–loud quasars, much larger than the gravitational radius ($R_g = 1.49 \times 10^{13} M_8$ cm, $M_8 = \frac{M_{bh}}{10^8 M_\odot}$), and even larger than the outer regions of the accretion disk. In the few sources in which the VLBI core is resolved (NGC1275, /2/) the observations revealed a complex structure without a single clearly collimated radio structure. As pointed out by Marscher in a recent review, "... the smallest emission region in a jet may be relatively far removed from the central engine...", and "...the association of non–thermal jets with black hole accretion remains a theoretical one." /3/. The observed super–luminal motion is usually interpreted in terms of relativistic speeds of the radio emitting plasma. Relativistic bulk velocities suggest that the motion originates from a very deep potential well but do not require it.

CGRO EGRET observations have shown that compact, flat spectrum, sources are also strong γ–ray emitters /4/. Most, and probably all, the EGRET Blazars exhibit super–luminal motion /4/, indicating that the jet is the source of the γ–ray emission. However, the γ–ray emission, is unlikely to come from very close to the central object, since the compactness of that region would be so high that photon-photon pair production should prevent most γ–ray from escaping the region. Therefore the variability observed in the γ–ray emission of 3C279 /5/ on time scales of days does not indicate a compact emission region (< 0.01 pc) but rather the effect of beaming. Again, γ–ray observations are probably not probing the innermost regions around the putative compact object, and a proof for the existence of a supermassive black hole based on detection of γ–ray emission is circumstantial.

From the variable luminosity and the time scale for its variation one can derive a minimum efficiency for the conversion of matter into radiation (e.g. Holt, this issue). If this exceeds $\sim 1\%$, then nuclear burning processes cannot be responsible. Early observations of the objects we now call Blazars, colored the debate by appearing to require enormous efficiencies so that a deep gravitational potential seemed the only way of producing the luminosity. It was soon clear however that this was rather the signature of relativistic beaming in these sources /6/. Leaving Blazars aside, we are left with a less impressive argument, as is shown in the following.

Microlensing

The most convincing examples of microlensing are the uncorrelated variations in the four images of the lens system Q2237+0305 ($z=1.695$, $L_{bol} \sim 3 \times 10^{46}$ erg s^{-1}, /7/). In this system the time delay between the four images is ~ 1 day, so that intrinsic variability can be distinguished from a genuine microlensing event. The limit to the dimension of the optical emission region (assumed homogeneous), obtained comparing the results of microlensing simulations to the observed variability events, is of $r_o \sim 2 \times 10^{15}$ cm /8/,/9/. This is a very compact size indeed, corresponding to only 70 R_g (if $\frac{L}{L_{Edd}} \sim 1$, which implies a black hole mass $M_8 = 2$. In the Starburst model of quasars the optical size of a Starburst–core of similar luminosity and at a similar redshift is measured in parsecs (even taking into account starburst mass segregation /10/), a factor at least 10^3 greater than the previous feature. Of course, one could try to explain the observed variability in terms of a few bright hot spots of size much smaller than r_o. In the Starburst model such hot spots could be compact supernova renmants (cSNR). Terlevich /10/, in fact, postulates that variability events of quasar are produced by cSNR, while most of the optical-UV (and bolometric) luminosity is provided by young stars. In Q2237+0305 no intrinsic variability (correlated in the four images) has however been detected, so cSNR cannot represent a significant fraction of the observed flux, unless the number of cSNR is large. In the latter case however, cSNR would provide most of the luminosity and this leads to the energetic problems pointed out by Heckman /11/ (because of the too short life of cSNR compared with even massive star lifetimes the total stellar mass implied to explain typical quasar luminosities, is 100–1000 times higher than the mass in the cores of present day elliptical galaxies, but see /10/). The number of cSNR should be not too low, in order not to produce appreciable intrinsic variability, but not too large, in order not to violate energetic constraints. It is clear that observations of microlensing in more than a single, maybe very special source, would create big trouble for the Starburst model and give support to the black hole hypothesis.

Variability in the X–ray Band

The largest and fastest variations of quasars are seen in the X–ray band. However, variability studies in this band are mainly confined to a number of bright nearby (and low luminosity) Seyfert galaxies. Probably the best, and most investigated, data set is that produced by EXOSAT. Green, McʰHardy & Letho, /12/, recently published a re–analysis of 110 light curves (all longer than 20 ksec) of 32 Seyfert galaxies and low-z quasars. They find that variability on time scales between 10 min and 1 day is common in these sources. *Ginga* observations of a few Seyfert galaxies push the limit on the variability time scale down to about 100 sec (e.g. NGC4051, MCG -6-30-15 /13/,/14/).

However, EXOSAT and *Ginga* observations are sometimes seriously source confused, as the dramatic case of NGC6814 shows well /15/. Until about one year ago NGC6814 represented the strongest and most celebrated case for the existence of a black hole in quasars. A stable periodicity, with factor 2 variations in one minute, and a strong correlation between the X–ray iron K_α line and the X–ray continuum, pointed to a highly compact system and suggested orbital clock. We now know that this is indeed true, but that these properties actually belong to a Galactic cataclysmic variable, 37 arcmin from the active nucleus /15/. Source confusion is clearly more a serious problem than was thought. Even in the highest X–ray flux sample of active galaxies /16/, about half have a source of comparable flux (factor of a few) closer than 1 degree (the *Ginga* field of view is $2° \times 1°$) seen in IPC and PSPC images. The standard confusion criteria developed for statistical logN–logS work are too weak to safeguard against confusion when one picks out a single peculiar object from a sample of order 50–100 objects. Instead of that standard 'one source in ~30' criterion one needs to require something much stronger. A reasonable safeguard would be to require a 1% chance that *any* of the 50–100 objects is strongly affected, i.e. 'one source in 5,000–10,000'. Much smaller beam sizes ($\lesssim 1$ arcmin for bright quasars) are then essential to be sure that confusion is not a problem. Presently long observations are few with imaging instruments. Rapid variables do still exist. Fig. 1 shows a 80 ksec ROSAT PSPC light curve of NGC4051. Large amplitude variations are clearly evident on time scales down to 100 sec in this low luminosity Seyfert galaxy. This would put a limit of $\approx 0.2\%$ on the efficiency for the conversion of matter into radiation. To clearly violate the nuclear power efficiency requires variations this short in 10–100 times higher luminosity quasars ($L_X \gtrsim 10^{43} - 10^{44}$ erg s^{-1}). EXOSAT found no variables at $L_X \gtrsim 10^{44}$ apart from 2 relativistically beamed objects.

Green et al, /12/, found a correlation between the normalized variability amplitude (NVA, the square root of the power at 2×10^{-4} Hz) and the X–ray luminosity for a sample of 12 quasars, in the sense that fainter sources show larger variations at the above frequency: NVA$\approx L_X^{-0.3}$. They interpret this result in terms of a inhomogeneous source where the overall emission comes from separate sub–regions which brighten randomly, and where the number of sub–regions is proportional to the size of the emission region. This is a "Christmas tree" model, as in the Starburst model. In the black hole model both the size of the emission region and its luminosity (assuming similar accretion rates) should scale with the mass of the black hole, giving a NVA$\propto L_x^{-0.5}$ relationship. One can get an exponent smaller than 0.5 by assuming that the sub–regions are not independent.

In the Starburst model the sub–regions are SN sites The typical time scales of the initial SN flashes are days to weeks, while that of the cSNR radiative phase is measured in months /10/. These kind

F. Fiore and M. Elvis

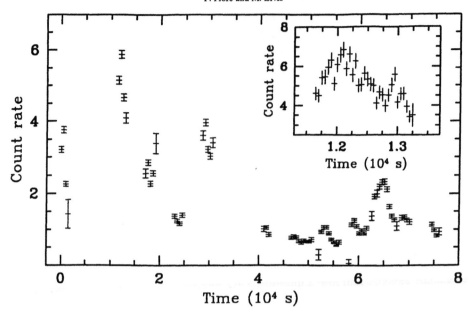

Fig. 1. ROSAT PSPC light curve of NGC4051 in bins of 500 s (50 s in the insert plot).

of events can explain the optical–UV variations seen in a few Seyfert galaxies like NGC4151 /10/, if the supernova rate is the order than 1 or a few supernovae per year. Rapid X–ray variability might be explained by the interaction of cSNR expanding shells with the dense and highly inhomogeneous ISM generated by previous SN explosions.

The Starburst model predicts a good correlation between the UV and X–ray variations (both mainly due to cSNR events) on a ~weeks time scale. Short time scale X–ray variations (due to the cSNR interactions with the ISM) should have no UV counterparts. Systematic simultaneous observations in optical, UV and X–rays have been performed for a handful of sources only, and the results show a complex situation. In NGC4051 there is no correlation between optical, UV and X–rays on short time scales (less than a night, /17/). In NGC4151 (and possibly NGC5548 /18/) the correlation between UV and X–rays on time scales of days to weeks is good when the source is in a faint UV state /19/, while when the UV is in a high state the correlation is absent /20/.

Detecting factor of 2 variations in luminous, unbeamed quasars would be a critical test of Starburst models, because if the supernova rate is high, as expected these objects, it is difficult to explain *any* variability on time scales of months. Although this experiment can be accomplished with ROSAT and ASCA, claims of variability in high luminosity quasars are limited to a few sources (Q0207–398 /21/, PKS2126–158 /22/ and S5 0014+813 /23/). In S5 0014+813 Elvis et al. /23/ found a ~ 30% increase in 7.2 months (rest frame). In the black hole model if this light travel time corresponds to ~ 20 R_g, than $r_X \lesssim 0.5$ pc, and the mass of the putative black hole is $M \lesssim 2 \times 10^{11} M_\odot$. The 4–30 keV luminosity of S5 0014+813 is about 10^{48} erg s^{-1} (assuming isotropic emission), requiring a mass ~ $10^{10} M_\odot$ in order not to violate the Eddington limit. This does not conflict with the variability limit but if the bolometric luminosity could be used, then the two limits would become very close. The above source size and luminosity would clearly exclude the Starburst model in this source (the size of a Starburst core of such luminosity is estimated to be hundreds of parsecs /10/). On the other hand the emission could be beamed, negating all these arguments. (Of course there would remain the difficulty of explaining the strong beaming in the Starburst model.)

3. GALACTIC ANALOGS

Jets

The strongest support for the black hole model using the existence of quasar jets comes from the similarity with the jets observed in a few galactic star systems, which almost certainly contain an accreting compact object, usually a black hole, like SS433, Cygnus X–3, 1E1740.7–2942 /24/, and especially GRS1915+105, where super–luminal motion has been recently detected /25/. On the other hand, Galactic super–luminal sources show that normal stellar evolution processes can lead to bulk relativistic motion. This add plausibility to the Starburst model but leaves unsolved the problem of aligning multiple jet sources.

Spectra

Several X–ray sources in our Galaxy and in the LMC show strong dynamical evidence that they contain black holes as part of a binary star system. The X–ray spectra of some low-z, broad emission line, quasars are similar to those of BHC in several ways: (a) spectra of both class of sources are (roughly) described by two components, a 'soft' component and a hard component; (b) The hard component in the band 1–50 keV is a power law with a remarkably small spread in energy index: $\alpha_E=0.5$–1.5; (c) Both quasar and BHC are not strong MeV emitters and there are a cases in both classes where a cut–off in the band 50–500 keV has been seen (/23/,/26/,/27/); (d) Both BHC and quasar exhibit broad edge–like structures at 7–10 keV /28/,/29/. Iron emission lines have been detected in sources of both classes with typical equivalent width 100 eV in Seyfert 1 and < 100 eV in BHC /28/,/29/,/30/.

The soft component of BHC can be parameterized in many ways, but, roughly speaking, it has a power law energy index of about 2–3. Also the soft component energy index of several low-z, radio–quiet quasars is between 2 and 3 /31/. The difference between the X–ray spectra of BHC and quasars is in the energy below which the soft component is dominant: about 1 keV in quasars /31/, 10 keV or even higher energies in BHC. Ebisawa and collaborators /28/,/32/ found indications that the soft component of BHC is generated close to the central compact object by fitting the same disk model (in which they take into account general relativity effects) to both BHC and neutron star sources. They found, systematically, a higher mass for the compact object in BHC (due to a lower observed maximum color temperature). By analogy, the even cooler soft component of quasars could be associated with the radiation coming from the innermost region of an accretion disk with a mass of $\approx 10^8 M_\odot$.

The time scales of variability for the soft and hard components of BHC are different. The soft component is roughly stable on time scales of 1 day or less, while the hard component exhibits large variations down to msec time scales /28/. If the time scales of BHC and quasars scale with the mass of the compact object the above two time scales translate in 10^4 years and 0.1 day respectively. Then, in a sample of quasars of similar z and luminosity one should find a rather small scatter in the soft component and a bigger scatter in the hard component. ROSAT results go exactly in this direction. Laor et al /33/ find that for that sample of 23 low-z PG quasars the scatter in the PSPC 2 keV luminosity is significantly larger than that in the 0.3 keV luminosity). This induces a correlation between the PSPC slope and the 2 keV X–ray quietness with respect to the optical (Fig. 2), in the sense that the hard component is generally fainter in the steeper sources.

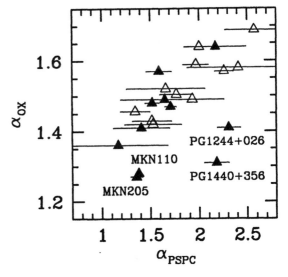

Fig. 2 The PSPC energy index plotted against the OUV to X–ray index α_{OX} for the quasars in /31/,/33/ and the PG objects in /35/. Open symbols are quasars with $M_B < -23.5$, filled symbols identify lower luminosity quasars.

BHC have two main states. BHC spend $\approx$ 10% of their time in the *high* state, with the relative intensity of the soft component much larger than in the *low* state. Do we see quasars that resemble BHC in the high (and soft) state? ROSAT finds that 10 % of quasars, in addition to BL Lacs /34/, have steep X–ray spectra ($\alpha_E \gtrsim 2$; $E < 2$ keV, /31/,/33/,/35/). There are sources, like MKN478 (PG1440+356), that have both steep X–ray spectrum *and* strong soft X–ray (and EUV) emission, i.e. they do not follow the Laor et al correlation (Fig. 2) /36/. These could be quasars in a 'high state'. It is not known whether these steep spectra extend to energies higher than 2 keV (the upper end of the PSPC range) nor if there is a hard tail. ASCA observations of these sources should answer these questions.

The first order similarity between BHC and quasar X–ray spectra argues for a black hole as the prime mover in the latter sources too. However the evidence is only circumstantial, not physics based, and so is relatively weak.

4. DYNAMICS: THE Fe–Kα LINE

If quasars are powered by accretion in a thin disk onto a supermassive black hole than spectral features will show signatures of orbital motions deep in the strong potential well of the black hole. These provide, in principle, strong tests of the black hole-accretion disk hypothesis and would allow measurements of the model parameters (black hole mass, disk inclination and geometry). No such features should be present in the Starburst model. The iron Kα line at 6.4 keV is especially important since it should come from the inner regions of an accretion disk ($\sim$ 50% from within 40 R_g /37/,/38/). The line ($0.3 < \sigma_\alpha < 0.8$ keV) will be broader than typical optical–UV BLR lines ($\sigma < 0.2$ keV). The simple detection of such broad iron Kα lines would be the best direct evidence of rotation, high velocities, and deep potential well in the innermost region ($< 100 R_g$) of quasars.

Matt et al. /38/ calculated the intensity and profile of this line from a flat, optically thick accretion disk rotating around a Schwarzschild black hole and illuminated by a central X–ray source, using a fully relativistic treatment and a Monte Carlo method for the radial and angular dependencies of the line emissivity. The line profiles are complex, often double horned and highly asymmetric. The details of the profile depends on the inclination, the inner and outer disk radii, the distance from the central X–ray source to the disk. At moderate inclination the asymmetry is largest, due to transverse and gravitational redshift. At high inclination ($\theta > 70°$) pure general relativity effects like gravitational focusing become important. This gives rise to the appearance of a rather strong structure between the two doppler horns /39/. Being a pure general relativity effect, its detection (and separation from any eventual narrow, low velocity component from larger radii) would be a strong proof of the black hole–accretion disk model.

Another strong test would come from studying the variations of the line profile in response to variations of the primary ionizing continuum /40/. This breaks the degeneracy that the line profile alone leaves between the mass of the central object and the size of the emission region. Stella /41/ showed that "bumps" will drift in frequency across the line profile toward the rest energy, as the illuminated region moves from the inner, high velocity regions to the outer, lower velocity ones. The drift time gives then the absolute radius of the disk and so the black hole mass. Matt & Perola /42/ proposed an analogous method which is less demanding in terms of energy resolution.

Unfortunately, X–ray observations of quasars have not, so far, produced a completely convincing detection of a broad iron Kα line. There have been claims for a $\sigma_\alpha \approx 0.3$ keV line in NGC5548 and IC4329A /43/, and a $\sigma_\alpha \approx 0.16$ keV line in MCG-6-30-15 /44/. On the other hand, a narrow line ($\sigma_\alpha \lesssim 0.1$ keV) has been detected in NGC7469 /45/. Detecting a broad iron line of equivalent width ~ 100 eV demands instruments with both good energy resolution and high sensitivity. EXOSAT and *Ginga* detectors had insufficient resolution and the ASCA GIS and SIS detectors may have insufficient sensitivity above 6 keV. A more robust conclusion must await deeper ASCA observations and/or the launch of the Spectrum X–γ, AXAF and XMM satellites.

4. EVOLUTION OF THE QUASAR SED

All the tests so far apply to individual quasars. In doing so they do not make use of one of the strongest properties of quasars, their strong evolution with redshift. The break point of the quasar luminosity function shifts to higher luminosity by a factor 40–50 between z=0.1 and z=2 /46/. This *population* evolution has been interpreted in terms of very different *physical* evolution models, such as, for example, black hole formation in the context of the hierarchical collapse of cold dark matter fluctuations /47/, or evolution of the star formation rate in the cores of elliptical galaxies /1/, with comparable levels of success. This means that *population* evolution is insufficient to distinguish even radically different quasar models. The reason for this probably resides on the virtually *scale-free* nature of quasar emission. From the very low luminosity quasars (i.e. 10^{40} erg s^{-1}) to the highest luminosity high–z quasars (10^{48} erg s^{-1} and maybe more), both continuum and emission lines scale almost linearly with luminosity, regardless of the redshift. Deviations from this patterns are small (e.g. the Baldwin effect). *Physical* evolution models could be more easily distinguished if there were characteristic luminosities or redshifts at which the properties of quasars changed. Thanks to new ROSAT and ASCA X-ray data on high–z quasars (/21/,/48/,/22/,/23/) "breaks" in the scaling laws for quasars are beginning to appear. These "breaks" can provide more stringent constraints on physical quasar models than population properties alone and hold the promise of providing strong tests of the black hole model.

Breaks in the Scaling Laws for Quasars

The first evidence for a change with z (or with the luminosity) of the X-ray spectrum of quasars may come from the observation that the difference between the X-ray spectrum of radio–quiet and radio–loud quasars *increases* with z (or with luminosity) /49/. We have expanded the original high–z sample in /49/ from a dozen sources to more than 40 by searching the ROSAT public archive for all known quasars with z> 1.5 and m_V <18. In Figures 3a,b we plot the hardness ratio R as a function of the Galactic N_H for radio–loud and radio–quiet sources respectively. [2]

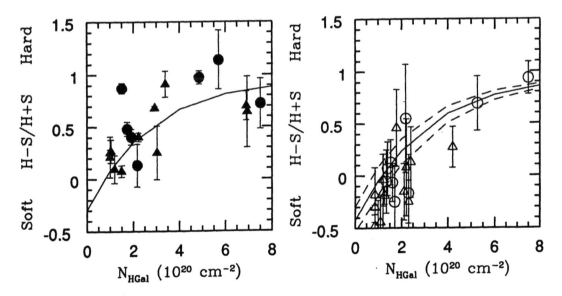

Fig. 3a. The PSPC hardness ratio for a sample of high–z, radio–loud quasars as a function of Galactic N_H. Circles identify z≈3 quasars while triangles identify z≈2 quasars.

Fig. 3b. Same as in Fig. 3a but for high–z, radio–quiet quasars.

[2] R=(H-S)/(H+S), H = counts in the 0.4-2.5 keV band, S = counts in the 0.1-0.4 keV band; note that the observed 0.1-2.5 band corresponds to 0.4-10 keV at z=3 and to 0.3-7.5 keV at z=2).

The solid (and dashed) lines show the expectation assuming the mean slope (and its dispersion) measured at low z in the 1–10 keV range /50/, /51/. Many of the radio–loud, high–z (and high luminosity) quasars are harder than their low–z (and low luminosity) counterparts. In some of the them this is due to heavy absorption /48/. However some (e.g. OQ172) show no evidence for absorption, but rather have intrinsically flat spectra. A *characteristic* luminosity or redshift may exist in radio–loud quasars. Conversely, we found no evidence of absorption in any of the few (4) radio–quiet quasars with enough counts to perform a proper spectral fit. Radio–quiet, high–z quasars appear to have an X–ray hardness consistent or *softer* than that of low–z quasars (Fig. 3b). The result holds even considering the data above 0.4 keV only (observed frame, 1.2 keV at z=2, Fig. 4). Some z=2 quasars are significantly softer, above 1.2 keV, than their low–z counterparts. If the softening is due to a contribution of an additional steep component entering the PSPC band, this component must extend to higher energies than for low–z quasars. In either case a *characteristic* redshift (or luminosity) may exist in radio–quiet quasars too. More data, especially at intermediate z and luminosity, are needed to confirm these trends.

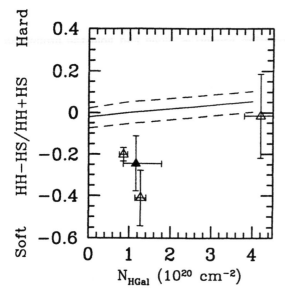

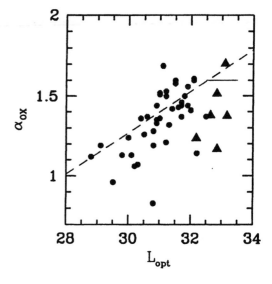

Fig. 4. The PSPC hardness ratio R=(HH–HS)/(HH+SS), HH=1.2–2.4 keV band, HS=0.4–1.2 keV band. The solid symbols identifies the average R of 4 quasars with similar N_{HGal}.

Fig. 5. α_{OX} versus log L_{opt} in ergs s^{-1} Hz^{-1}, for steep spectrum, radio–loud quasars. Circles are quasars detected with the *Einstein* IPC, triangles are z≈3 quasars reported in /21/ and /49/.

The second piece of evidence of a change with luminosity (or with z) of the quasar continuum comes from X–ray loudness. Fig. 5 is a plot of the X–ray loudness α_{OX} as a function of the optical luminosity L_{opt} for those radio–loud quasars in which beaming is not likely to be important (steep–spectrum and GigaHertz Peaked radio sources /52/). The dashed line shows the predicted relation using for steep radio spectrum quasars /53/. The horizontal solid line represents the prediction for Compact Steep Spectrum sources, the objects most similar to our high–z quasars (which are mostly Gigahertz peaked). The high–z quasars lie systematically below the predictions. They are too X–ray loud for their optical luminosity. This may possibly be the result of a selection effect, since some of the high z quasars were selected because their known X–ray or radio brightness. Still the break in the predictive power of α_{OX}, may indicate a characteristic luminosity at $\log L_{opt} \sim 32 - 33$ for radio–loud, compact, non–flat spectrum quasars.

Toy Models for Characteristic Quasar Luminosities

Models for the *physical* evolution of quasars must accommodate the above results about the existence of characteristics redshift and luminosities. This means that they need to incorporate both models for the generation of the X–ray luminosity, and for the emission mechanisms at work. *Toy models* of this kind, in the framework of the black hole model, are as follows (also see /49/):

(1) There are two extreme possibilities for the quasar physical evolution in the black hole model: individual objects are long–lived, or short–lived /54/. If quasars are numerous, short–lived and result from the collapsing CDM halos in the Haehnelt & Rees picture, then a typical black hole mass will *decrease* with z. On the other hand, if quasars were rare and long–lived, then the black hole masses *increase* with decreasing z.

Assume that comptonization of soft photons by a population of thermal electrons in a hot corona above a layer of cold reflecting matter produces the 2–10 keV spectrum in X–ray quiet quasars /55/. In this model accretion onto large black holes ($\approx 10^{10} M_\odot$ for these luminous, z=2–3 quasars) results in steeper 2–10 keV spectra than accretion onto small black holes. So short–lived models will produce steeper spectra at high z, in agreement with our findings. This holds for a range of corona optical depths, and therefore luminosities, in units of the L_{Edd} /55/.

(2) How can a characteristic luminosity in unbeamed radio–loud quasars be interpreted? There are two potential sources of energy in a massive black hole model: accretion and spin /56/. The luminosity due to the extraction of the gravitational energy of accreting gas L_{accr} has a linear dependence on the black hole mass:

$$L_{accr} = 1.28 \times 10^{46} \frac{L_{accr}}{L_{Edd}} M_8 \ erg \ s^{-1}.$$

We may identify L_{accr} with the optical–UV luminosity, with the X–ray luminosity of radio–quiet quasars and with some part of the X–ray luminosity of radio–loud quasars, say a fraction f of L_{accr}. On the other hand, the electromagnetic luminosity, L_{em}, given by Blandford /56/, has a different dependence on the black hole mass than L_{accr}:

$$L_{em} \approx 10^{45} (\frac{a}{m})^2 B_4^2 \ M_8^2 \ erg \ s^{-1},$$

where B is the magnetic field in units of 10^4 G, a is the specific angular momentum and $m = GM/c^2$. The X–ray luminosity in radio–loud quasars could be proportional to L_{em}, say a fraction g of L_{em}. Comparing the above two equations then defines a critical mass for radio–loud quasars at which L_{Xem} equals L_{Xaccr}. We can interpret the critical luminosity of $\log L_{opt} \sim 32 - 33$ erg s^{-1} Hz^{-1} as corresponding to this critical mass for radio–loud quasars. For an Eddington limited source this corresponds to a black hole mass of $10^9 - 10^{10} M_\odot$, since the luminosity per decade, $L_{opt} = 10^{47} - 10^{48}$ erg $^{-1}$ (see /49/).

These toy models, although not representing yet a test for the black hole model, show how promising and how still unexplored are the possibilities for studying the evolution of the quasar continuum.

ACKNOWLEDGMENTS

J. Bechtold, A. Laor, J. McDowell, A. Siemiginowska and B. Wilkes have all contributed to the studies presented in this paper. We also thanks G.C. Perola and G. Matt for useful discussions.

REFERENCES

1. R.J. Terlevich, & B.J. Boyle, *M.N.R.A.S.*, 262, 491 (1993).
2. T. Venturi, A.C.S. Readhead, J.M. Marr, & D.C. Backer, *Ap. J.*, 411, 522 (1993).
3. A.P. Marscher, in: *Testing the AGN Paradigm*, AIP Conference Proceedings 254, 1992, p.377.
4. R.C. Hartman, et al., in: *The Second Compton Symposium*, AIP Conference Proceedings 304, 1993, p.563.
5. D.A. Kniffen, et al., *Ap. J.* 411, 133 (1993).
6. G.C. Perola, *Physica Scripta*, T7, 142 (1984).
7. M.J. Irwin, R.L. Webster, P.C. Hewett, R.T. Corrigan, & R.I. Jedrzejewski, *A. J.*, 98(6), 1989 (1989).

8. J. Wambsganss, B. Paczynski, & P. Schneider, *Ap. J. Letters*, 358, L33 (1990).

9. K.P. Rauch, & R.D. Blandford, *Ap J. Letters*, 381, L39 (1991).

10. R.J. Terlevich, in: *Violent Star Formation from 30 Doradus to QSOs*, Puerto Naos, 1994.

11. T.M. Heckman, in: *Mass-transfer-induced Activity in Galaxies*, University of Kentucky, 1994.

12. A.R. Green, I.M. McHardy, & H.J. Letho, *M.N.R.A.S.*, 265. 664 (1993).

13. M. Matsuoka, L. Piro, M. Yamauchi, & T. Murakami, *Ap. J.*, 361, 440 (1990).

14. F. Fiore, G.C. Perola, M. Matsuoka, & M. Yamauchi, *A.A.*, 262, 37 (1992). 15. G. Madejski, et al., *Nature*, 365, 626 (1993).

16. G. Piccinotti, et al., *Ap. J*, 263, 485 (1982).

17. C. Done, et al., *M.N.R.A.S.*, 243, 713 (1990).

18. J. Clavel, et al., *Ap. J.*, 393, 113 (1992).

19. G.C. Perola, et al., *Ap. J.*, (1986).

20. G.C. Perola, & L. Piro, *A. A.*, (1994).

21. J. Bechtold, et al., *A. J.*, 108(2), 374 (1994a).

22. P. Selermitsos, et al., *Ap. J. Letters*, in press (1994).

23. M. Elvis, M. Matsuoka, A. Siemiginowska, F. Fiore, T. Mihara, & W. Brinkmann, *Ap. J. Letters*, 435 in press (1994b).

24. I.F. Mirabel, L.F. Rodriguez, B. Cordier, J. Paul, & F. Lebrun, *Nature*, 358, 215 (1992)

25. I.F. Mirabel, & L.F. Rodriguez, *Nature*, 371, 46 (1994)

26. M. Maisak, et al., *Ap. J. Letters*, 407, L61 (1993).

27. E.P. Liang in: *Compton Gamma-ray Observatory*, AIP Conference Proceedings 280, 1993, p. 396.

28. K. Ebisawa, Ph. D. Thesis, University of Tokyo 1991.

29. K.A. Pounds, K. Nandra, G.C. Stewart, I.M. George, & A.C. Fabian, *Nature*, 344, 132 (1990).

30. K. Nandra, & K.A. Pounds, *M.N.R.A.S.*, in press (1994).

31. F. Fiore, E. Elvis, A. Siemiginowska, B.J. Wilkes, & J.C. McDowell, *Ap. J.*, 431, 515.

32. K. Ebisawa, K. Mitsuda, & T. Hanawa, *Ap. J.*, 367, 213 (1991).

33. A. Laor, F. Fiore, M. Elvis, B.J. Wilkes, & J.C. McDowell, *Ap. J.*, 435 in press (1994).

34. N.E. White, A.C. Fabian & R. F. Mushotzky, *A. A. Letters*, 133, L9 (1984).

35. R. Walter, & H.H. Fink, *A. A.*, 274, 105 1993.

36. F. Fiore, E. Elvis, A. Siemiginowska, B.J. Wilkes, J.C. McDowell, & S. Mathur *Ap. J.*, submitted (1994).

37. G. Matt, G.C. Perola, & L. Piro, *A. A.*, 247, 25 (1991)..

38. G. Matt, G.C. Perola, L. Piro, & L. Stella, *A. A.*, 257, 63 (1992).

39. G. Matt, G.C. Perola, & L. Stella, *A. A.*, 267, 643 (1993).

40. R.D. Blandford, & C.F. McKee, *Ap. J.*, 255, 419 (1982).

41. L. Stella, *Nature*, 344, 747 (1990).

42. G. Matt, & G.C Perola, *M.N.R.A.S.*, 259, 433 (1992).

43. R. Mushotzky, et al., *M.N.R.A.S.*, submitted (1994).

44. A.C Fabian, et al., *P.A.S.J.*, 46, L59, (1994).

45. L. Piro, private communication (1994).

46. B.J. Boyle, R. Fong, & T. Shanks, *M.N.R.A.S.*, 227, 717 (1987).

47. M.G. Haehnelt, & M.J Rees, *M.N.R.A.S.*, 263, 168 (1993).

48. M. Elvis, F. Fiore, B.J. Wilkes, J.C. McDowell, & J. Bechtold, *Ap. J.*, 422, 60 (1994).

49. J. Bechtold, et al., *A. J.*, 108, 759 (1994b).

50 A.J. Lawson, M. Turner, O.R. Williams, G.C. Stewart & R.D. Saxton, *M.N.R.A.S*, 259, 743 (1992).

51. O.R. Williams, et al., *Ap. J.*, 389, 157 (1992).

52 C.P. O'Dea, S.A. Baum, & C. Stanghellini, *Ap. J.*, 380, 66 (1991).

53. D.M. Worrall, P. Giommi, H. Tananbaum, & G. Zamorani *Ap. J.*, 313, 596 (1987).

54. A. Cavaliere, & P. Padovani, *Ap. J.*, 315, 411 (1988).

55. F. Haardt, & L. Maraschi, *Ap. J.*, 413, 507 (1993).

56. R.D. Blandford in: *Active Galactic Nuclei*, eds. T.J.-L. Courvoisier and M. Mayor, Springer-Verlag, Berlin 1990.

Adv. Space Res. Vol. 19, No. 1, pp. (1)95–(1)98, 1997
© 1997 COSPAR
Printed in Great Britain. All rights reserved
0273–1177/97 $17.00 + 0.00

 Pergamon

PII: S0273-1177(97)00042-2

ASCA OBSERVATIONS OF SEYFERT 2 GALAXIES

H. Awaki*, S. Ueno*, K. Koyama*, K. Iwasawa**,
H. Kunieda** and ASCA team

* *Department of Physics, Kyoto University, Kitashirakawa, Sakyo,
Kyoto 606-01, Japan*
** *Department of Physics, Nagoya University, Furo-cho, Chikusa, Nagoya
466-01, Japan*

ABSTRACT

The Japanese X-ray satellite, ASCA has observed 6 Seyfert 2 galaxies. We found the evidence of obscured nuclei in all galaxies, i.e. heavy absorption feature for four galaxies and strong lines probably produced by fluorescence for two galaxies. This result supports the unified scheme of AGNs, and is very important to reveal the nature of type 2 Seyferts, e.g. obscured nucleus, scattering region, etc. The thermal component is found in low energy band as well as non thermal emission for two galaxies, and it would originate from starburst activity. It is very important, considering the connection between AGN and Starburst galaxy. It may give us a hint of the galaxy evolution.

INTRODUCTION

Seyfert 2 galaxy is one class of AGNs, and it had been considered as less active AGN because of the lack of broad lines in optical spectrum. Many Seyfert 2 galaxies were observed by Einstein satellite, and it is found that the X-ray luminosity of Seyfert 2 galaxy was less than that of Seyfert 1 galaxy /1/. This result confirmed less activity of Seyfert 2 nuclei, and it seemed to be inconsistent with the unified scheme of Seyfert galaxy /2/. Using high sensitivity of Ginga in the 2-20 keV band, heavily obscured nuclei have been discovered in several Seyfert 2 galaxies /3/,/4/,/5/. The obscured column is about $10^{23} cm^{-2}$. This supports the unified scheme, and the less activity in the Einstein band is naturally explained by the extinction of the nuclear emission due to the obscured matter.

Type 2 Seyfert or obscured nuclei of galaxies become one of the important issue of X-ray astronomy these days, because of its strong impact to the unified scheme of AGN. Although Ginga found an obscured nucleus in Seyfert 2 galaxy, a circumnuclear region including the accretion torus is still unknown. ASCA has spectrometers with the energy band from the 0.3 keV to 10 keV, and we can see both X-ray characteristics observed in the Einstein and the Ginga bands. We will reveal the circumnuclear region using ASCA. Seyfert 2 galaxies are good candidates to examine a connection between Starburst galaxy and AGN /6/.

ASCA OBSERVATIONS

ASCA carries imaging spectrometers with high energy resolution, and has a large effective area in the wide energy band from 0.3 to 10 keV/7/. We observed six Seyfert 2 galaxies, NGC1068, Mkn3, NGC4507, NGC4388, NGC4945, and IC5063, for which obscured nuclei have been reported /5/,/8/. The data was taken with both SIS and GIS detectors, and the exposure time was about 40ksec for each galaxy. The SIS data was only used for our analysis, because the SIS has better energy resolution and high sensitivity in soft band. The standard analysis was performed on the obtained data, and then got an

X-ray spectrum. The obtained X-ray spectra are classified into that with heavy absorption feature and that with strong iron lines.

Heavily Obscured Nucleus

Fig.1 is an X-ray spectrum of NGC4507 which has a typical spectrum in this category. The spectrum cannot be described as single component model. Therefore we fitted it with two components model, which is *wabs1(powl + wabs2(powl+gauss))* in XSPEC expression, where *wabs , powl* and *gauss* mean photo absorption in cold matter, a power low continuum, and a gauss function, respectively. We fixed the column density in *wabs1* to be the Galactic column. Since it is difficult to determine the photon index of soft component, we fixed it the same value of hard one. This model is also acceptable. Table 1 shows the best fit value. We note that the soft and the absorbed components, which are detected with Einstein and Ginga, separately, are detected with ASCA.

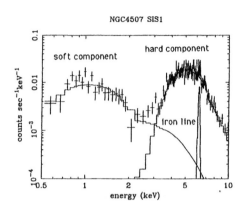

Fig.1 The X-ray spectrum for NGC4507 with the ASCA SIS.

TABLE 1 Observational Results for Seyfert 2 Galaxies with Heavy Absorption Feature.

NAME	Lx (2-10keV)	Γ_1	Γ_2	N_H x10^{23} cm^{-2}	EW_narrow (eV)	References
Mkn3	7.2×10^{42}	$= \Gamma_2$	1.78(1.63-1.95)	4.42(3.32-5.77)	600(390-910)	9
NGC4507	4.4×10^{43}	$= \Gamma_2$	1.79(1.65-1.93)	3.66(3.50-3.82)	110(50-170)	10
NGC4388	3.9×10^{42}	$= \Gamma_2$	1.19(0.81-1.57)	3.81(3.29-4.34)	380(200-550)	10
IC5063	2.1×10^{43}	$= \Gamma_2$	1.89(1.75-2.04)	2.47(2.34-2.60)	60(10-110)	10

( ) : 90% confidence region

The X-ray luminosity and photon index of the absorbed continuum emission for each galaxy are comparable to those of Seyfert 1 galaxy /11/. This result indicates the existence of a luminous nucleus, which is obscured by thick matter with the hydrogen column density of about 10^{23}cm^{-2}. It is consistent with a prediction from Unified Seyfert model. Considering that the photon index of the soft component is the same as that of the hard one, the ratio of the soft component to the hard component are 10% for Mkn3, 0.67% for NGC4507 and IC5063, and 2.3% for NGC4388. Higher ratio for Mkn3 and NGC4388 are caused by the time variation of the hard component and the contamination of the diffuse emission filled in Virgo cluster, respectively. Mulchaey et al. mentioned the ratio range from 1% to 10% using Einstein and Ginga data/12/. However they compared the data by different instruments. The intensity of the soft component is almost equal between Einstein and ASCA observations, and those galaxies are less starburst activity. Therefore, the soft component would be a scattered light of the nuclear emission.

Iron line emission is also found from each galaxy. The equivalent width ranges from 100 to 600 eV, which is larger than that of Seyfert 1 galaxy /5/. Since the iron line would be produced through fluorescence, the intensity of iron line trace total amount of a matter around the nucleus. In the unified model, the matter corresponds to an accretion torus. The estimated covering factor of torus are > 70%, for Mkn3 and NGC4388, ~ 50% for NGC4507, and ~40% for IC5063. We note that the ratio is not 100% because of the detection of the scattered light. These are almost consistent with the observation in other band /13/. In detail analysis, we found the line broadening of 0.094(0.053-0.137) keV for Mkn3. The line width corresponds to 10 thousands km/s, and it is similar to the width of broad lines in optical band. For NGC4507, NGC4388, and IC5063, we got only upper limit of about 0.2 keV.

We can not detect any significant time variation during our observation, but find the long term variation for Mkn 3. This galaxy has been observed with various experiments several times. The soft component is stable from *Einstein* and *ASCA* observations. It indicates that the scattering region would be greater than 3pc. On the other hand, the hard component is decreasing during *Ginga - BBXRT - ASCA* observations, and the intensity of iron follows the decrease of hard component. The iron line emitting region would be less than 0.6pc.

AGN with Strong Iron Line

We found the strong iron line in the spectra of NGC1068 and the central region of NGC4945. The spectra are shown in Fig. 2. We fit the spectra above 3 keV with single power law plus a line. The equivalent width of the line is greater than 1 keV. Since the strong line associate with the scattered light, the observed light would be a portion of the nuclear emission through the scattering process. Applying the ratio between scattered light and direct light in above section, the intrinsic luminosity for NGC1068 and NGC4945 are to be 3×10^{43} and 6×10^{41} ergs/s, respectively. They are similar to those for Seyfert 1 galaxies. For NGC4945, our estimation is consistent with the intrinsic luminosity of 10^{42} ergs/s observed by Ginga /14/.

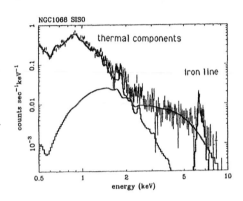 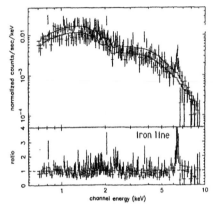

Fig. 2 The ASCA SIS spectra for NGC1068 (left) and a core region of NGC4945 (right).

TABLE 2 Results for the Galaxies with Strong Iron lines

	NGC1068/15/	NGC4945/16/
Lx (2-10keV)	2.9×10^{41}	6×10^{39}
Γ	1.27 ± 0.3	1.1
N_H ($\times 10^{21}$ cm^{-2})	< 50	0.7 < 5
E1 (keV)	6.38 ± 0.01	6.4
EW1 (keV)	1.6 ± 0.4 (Narrow)	1.2 ($\sigma \sim 0.15$ keV)
E2 (keV)	6.60 ± 0.03	6.7 (fixed)
EW2 (keV)	1.0 ± 0.5 (Narrow)	< 0.1
E3 (keV)	6.85 ± 0.02	6.9 (fixed)
EW3 (keV)	0.6 ± 0.4 (Narrow)	< 0.1

I(E) = wabs (powl+ gauss + gauss ...) in XSPEC expression.

For NGC1068, the line can be separated into three narrow lines., i.e. a neutral line, a He-like line, a H-like line of iron. A neutral line is associated with cold/warm medium and a H-like line associated with hot medium. There would be multi-components in the scattering region. Marshall et al. assumes two components, which are warm component with the temperature of 2×10^5K and hot component with the temperature of 4×10^6K /17/. For NGC4945, we didn't detect highly ionized iron line. It indicates the absence of hot component.

Connection with Starburst Galaxy

It is known that Seyfert 2 galaxies often have starburst activity as well as Seyfert activity. NGC1068 and NGC4945 also have starburst activity. These galaxies are very important, considering the connection of AGN with starburst galaxies. Fig2a and Fig 3 are the spectra for NGC1068 and NGC4945.

For NGC1068, the X-ray emission below 3keV cannot be explained by the extrapolation of high energy component. A soft component is required. The soft component is well described as thermal emissions with different temperatures of 0.59 keV and 0.14 keV /15/. Since an extended structure with size of about 3kpc has been found with ROSAT /18/, the hot gas would be extended similar to a starburst galaxy. We estimate the luminosity of the thermal component, and found that our results are similar to those for Starburst galaxy, e.g. the same correlation between a soft X-ray luminosity and an infrared luminosity /19/ or Br$_\gamma$ luminosity /20/ as those for starburst galaxy. The thermal emission would come from circumnuclear starburst region. NGC4945 has an extended X-ray emission with ~5 arcmin. We make the spectrum in north-east region of NGC4945, and then fit the spectrum with thermal or power law emission. The fitting result is listed in Table 3. Because of a detection of a He like Iron line, the continuum emission would include a thermal emission with temperature of several keV. Since the temperature is higher than that of starburst galaxies, and the extended emission is aligned the galactic disk, the emission might be a kind of the Galactic ridge emission.

TABLE 3 Characteristics of diffuse X-ray emission

	NGC1068	NGC4945
Lx (0.5-4.5)*	4.7×10^{41}	$\sim 3 \times 10^{39}$
kT1 (keV)	0.14±0.1	~6
Z1	0.03±0.01	1(fixed)
kT2 (keV)	0.59±0.1	
Z2	0.30±0.06	

*: only thermal components

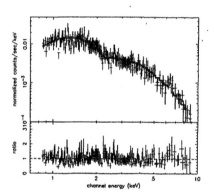

Fig.3. The X-ray spectrum for the extended region of NGC4945

REFERENCES

1. G.A.Kriss et al. Astrophys. J., 242, 492-501 (1980)
2. R.R.J.Antonucci, and J.S.Miller, Astrophys. J., 297, 621-632 (1985)
3. K.Koyama et al., Publ. Astron. Soc. Japan, 41, 731-737 (1989)
4. R.S.Warwick et al., Publ. Astron. Soc. Japan, 41, 739-744 (1989)
5. H. Awaki et al., Publ. Astron. Soc. Japan, 43, 195-212 (1991)
6. Heckman, IAU syposium, No.134, p357-364 (1988)
7. Y. Tanaka, H. Inoue, and S.S. Holt, Publ. Astron. Soc. Japan, 46, L37-L41 (1994)
8. C.G. Hanson et al. Monthly Notice Roy. Astron. Soc., 242, 262-266 (1990)
9. K.Iwasawa et al., Publ. Astron. Soc. Japan, in press (1994)
10. S. Ueno, private communication (1994)
11. J. Turner and K.A. Punds, Monthly Notice Roy. Astron. Soc., 240, 833-880 (1989)
12. Mulchaey et al. Astrophys. J., 390, L69-L72 (1992)
13. A.Wilson et al., Astronomical J., 107 .1227-1234 (1994)
14. K. Iwasawa et al. Astrophys. J., 409, 155-161 (1993)
15. S. Ueno et al., Publ. Astron. Soc. Japan 46, L71-L75 (1994)
16. K. Iwasawa, private communication (1994)
17. F. Marshall et al. Astrophys. J., 405, 168- 178 (1993)
18. A.S. Wilson et al., Astrophys. J., 391, L75- L79 (1992)
19. L.P. David, C. Jones, and W. Forman, Astrophys. J., 388, 82-92 (1992)
20. M. Ward Monthly Notice Roy. Astron. Soc., 231, 1p-5p (1988)

Adv. Space Res. Vol. 19, No. 1, pp. (1)99–(1)108, 1997
Published by Elsevier Science Ltd on behalf of COSPAR
Printed in Great Britain
0273–1177/97 $17.00 + 0.00

PII: S0273-1177(97)00043-4

COMPTONIZATION MODELS AND SPECTROSCOPY OF X-RAY AND GAMMA-RAY SOURCES

Lev G. Titarchuk* and Xin-Min Hua**

* *NASA Goddard Space Flight Center, Code 668, Greenbelt, MD 20771, U.S.A.*
** *ISTS North York, Ontario M3J 3K1, Canada*

ABSTRACT

We compare the analytical Generalized Comptonization model which takes into account the relativistic effects /1/, and Monte Carlo calculations for photon Comptonization by relativistic plasma clouds. We show that the new analytical model extends the previous work to a much wider range of plasma temperatures and optical depths. In general, the emergent spectra from a hot plasma cloud depend upon the spectral and spatial distributions of source photons as well as the plasma temperature and geometry. Based on the comparison between the theoretical and Monte Carlo calculations, we determine quantitatively a range of plasma geometry parameters and temperatures for which the emergent upscattering spectra are insensitive to the spectral and spatial distribution of source photons. Within this parameter range, we show that the shape of the emergent spectrum depends on two parameters only, namely the plasma temperature and β, the parameter which characterizes the photon distribution over the number of scatterings which the soft photons undergo in the plasma cloud in order to become the hard ones. We find the exact solution of the Kompaneets equation in the case of subrelativistic energies and plasma temperatures, in the optically thick regime. The solution recovers the low and high energy asymptotic forms studied in /1/. Also, we modify the Sunyaev & Titarchuk formula /2/ in order to obtain a fairly good subrealtivistic analytical approximation. This new formula, verified by Monte-Carlo calculations, contains as partial cases the Titarchuk's analytical results /1/, for optically thick and thin cases respectively. The analytical models, examined by the present Monte Carlo calculations, make possible a more efficient spectral analysis of data obtained from X-ray and gamma-ray sources.
Published by Elsevier Science on behalf of COSPAR

I. INTRODUCTION AND SUMMARY

In previous work, the analysis of Compton scattering in hot electron plasma was simplified by assuming a non-relativistic approximation and by considering limited ranges of photon and electron energies. The problem was also dealt with using numerical techniques such as the Monte Carlo method (see e.g. /3/). The Comptonization model /2/ (hereafter ST80), presented the analytical form of the emergent spectra obtained in the diffusion approximation of photon scatterings in plasma clouds with optical depths $\tau_0 \gg 1$, with the assumptions of nonrelativistic photon energy and plasma temperature. A proper description of photon scatterings in plasma clouds of arbitrary optical depths and relativistic temperatures, however, requires relativistic corrections, or even a fully relativistic treatment. Several papers have been devoted to this issue /4-9/.

Titarchuk (/1/, hereafter T94) generalized the ST80 model, and presented the analytical theory of X-ray and gamma-ray spectral formation in plasma clouds with small as well as large optical depths where relativistic effects and Klein-Nishina corrections are important. In this study, we apply an efficient Monte Carlo technique which deals with the Comptonization problem in a fully

†NAS–NRC Research Associate

relativistic way. By comparison between the analytical and Monte Carlo calculations, we show that the Generalized Comptonization models of T94 are applicable to a wide range of plasma temperatures and cloud optical depths. With the help of Monte Carlo calculations, we make a detailed study of the conditions for the analytical models T94 to be applicable.

As was emphasized in /2/,/10/ (hereafter ST85), and T94, the problem of the X-ray spectral formation in a hot plasma cloud is closely connected to the distribution of the number of scatterings u that photons undergo before escaping. If the plasma temperature kT_e is subrelativistic, i.e. $kT_e \lesssim 500$ keV, and characteristic energy of the source photons $kT_r \ll kT_e$, then the photons have to undergo a large number of scatterings (by diffusion in the plasma cloud), in order to reach subrelativistic energies, i.e. $h\nu \sim kT_e$. If this number of scatterings u is much greater than the average number $\bar{u}$, the hard photon distribution $P(u)$ over scattering number u follows the exponential law of Eq.(II.2), with the characteristic scale $1/\beta$, where β is uniquely determined by the Thomson optical depth τ_0 of the plasma cloud and the geometry (Eq. II.3). β is independent of the source photon spectrum and spatial distribution.

In practical use, the analytical solutions Eqs.[(II.6), (II.10) and (II.11)] are applicable to the situation where the emergent spectra are not sensitive to the source conditions. We describe these conditions in terms of kT_e, β and the source radiation temperature T_r by means of the following inequality: $y \approx kT_e/(\beta m_e c^2) \ll (1/6) \ln (T_e/T_r)$ /11/. In other words, the unsaturated Comptonization regime is required (Comptonization parameter y of order unity). By comparison with the Monte Carlo results, we are able to give a more quantitative determination of the parameter range where these conditions are met (see Fig.3 and discussion in §III). We confirm the prediction of T94 that in this range the shapes of emergent spectra depend only on temperature and β, irrespective of the geometry of clouds. In the general case, such as in the case of saturated Comptonization, the emergent spectra should be represented by linear superpositions of the appropriate fundamental spectra with corresponding Fourier coefficients $\{c_i\}$ obtained from the spatial distribution of source photons /1/.

In §II.i we find the exact solution of the Kompaneets equation Eq.(II.6) in the case of subrelativistic energies and plasma temperatures in the diffusion regime ($\tau_0 > 1$). The solution recovers the low and high energy spectral asymptotic forms studied in /1/. Also we modify the ST80 formula (§II.ii) in order to produce a fairly good subrealtivistic analytical approximation. The Compton scattering cross section drops substantially when the electron temperature and the photon energy go up (see e.g. /11/). In other words, a cloud is more transparent to higher energy photons and less capable of changing their energies and directions. Consequently, even in plasma clouds with τ_0 fairly large and $kT_e > 100$ keV, most of the photons do not reach the Wien barrier $3kT_e$, and the Wien hard tail $x^3 \exp(-x)$ is more or less suppressed. Quantitatively this effect could be accounted for by introducing the effective optical depth τ_R for photons at the hard tail ($E > kT_e$) and by the appropriate reducing factor r. This new formula Eq.(II.9), verified by Monte-Carlo calculations, includes as partial cases all the T94 analytical results for optically thick and thin cases respectively.

In §II.iii we present the analytical spectra obtained as a result of the upscattering of the initial soft photons to the very high energies, in the case of moderate depths ($\tau_0 \lesssim 1$). In order to gain energy, the primary photons have to undergo a certain number of collisions k in the hot subrelativistic electron gas. A soft photon of energy E_1 attains an energy E_2,

$$E_2 = E_1 \frac{c - v \cos \theta_1}{c - v \cos \theta_2}$$

when scattering off an electron moving with velocity v. $\theta_{1,2}$ are the angles between the electron velocity and the direction of the photon propagation (before and after scattering respectively). The energy gain is maximal when the electron moves towards the photon, and scatters the photon backwards, $\theta_1 = -\pi$, $\theta_2 = 0$. The last condition is not very restrictive because even mildly relativistic electrons scatter photons predominantly in the direction of their motion. The dependence of the scattering probability on θ_1 is not strong, $\propto 1 - \frac{v}{c}\cos\theta_1$, however we are not interested in the scattering of every photon, but only in that small fraction which can reach high energies. In a medium with moderate optical depths, the probability of the photon to be scattered k times decreases rapidly with k. Therefore only photons that scattered effectively enough, i.e. $\theta_1 \approx -\pi$, $\theta_2 \approx 0$, contribute to the hard tail of the spectrum because they get the required energy after a minimal number of scatterings. Photons which scattered ineffectively escape before they reach high energy. *So photons in the hard tail may be considered as scattering only in the backward and forward direction.* This implies a specific angular distribution of the emergent radiation, as seen by an observer on the surface of the plasma cloud. In a disk geometry the effective propagation of those high energy photons is predominatly in the plane of the disk because photons propagating at a large angle to this plane escapes readily. Similarly in a spherical geometry photons scattered backward and forward a sufficiently large number of times should propagate close to the diameters of the sphere. For example, in the case of electron temperature 250 keV and Thomson optical radius 0.2, our Monte Carlo calculations show that the intensity of the radiation emergent at an angle $\theta = 30^0$ is 6 times less than the intensity along the radial direction ($\theta = 0^0$), in agreement with the above arguments. This specific angular distribution determines the dependence of the Comptonization parameter $y = 1/\gamma_0$ (or the spectral index α_0) on the electron temperature [Eq.(II.10) and see also Appendix B in T94]. The analytical spectra in the optically moderate case are presented by the equations (II.9-11).

The analytical formula of the emergent spectrum is given for a Wien source injection in §II.**iv**. The effect of the non-monochromatic source soft photon injection on the emergent upscattering spectra is considered. As it is well known, the high energy tail is independent of the spectrum of the soft photon source.

The results of comparisons of Monte Carlo and analytical calculations are discussed in §III. Figures 1,2 illustrate good agreement between Monte Carlo calculated spectra and analytical spectra, in a wide range of electron temperature and the optical depth values. The applicability region of the analytical technique in terms of the parameters kT_e and β is also discussed in §III, and is displayed in Fig.3.

II. THE COMPTONIZATION MODELS

As is well known, for non-relativistic scattering of photons by a thermal plasma of temperature T_e, the mean change in photon frequency is given by $\delta\nu/\nu = (4kT_e - h\nu)/m_ec^2$. The shape of the radiation spectrum is determined solely by the Comptonization parameter $y \approx kT_e/m_ec^2\bar{u}$. Here $\bar{u}$ is the average number of scatterings which photons suffer in the plasma cloud of Thomson optical depth τ_0. In /1/ and early in /2/, we pointed out several times that for the upscattering Comptonization, only the number of scatterings of those initial photons which became hard ones is important. *This number of scattering $\bar{u}$ determines the Comptonization parameter y.* The spatial source distribution determines a specific number of scatterings which photons undergo in plasma cloud. In our upscattering case it is related with the first spatial distribution eigenfunction of the appropriate diffusion problem /1/. For example, the scattering number being equal to $\tau_0^2/5$ (the scattering number *for all photons* which fill uniformly a sphere, ST80) is irrelevant in the upscattering case.

In the diffuse approximation ($\tau_0 \gg 1$) for the spherical case, $\bar{u} \approx 3\tau_0^2/\pi^2$ (e.g. ST80) and thus

$$y \approx \frac{kT_e}{m_ec^2}\frac{3\tau_0^2}{\pi^2}. \qquad (II.1)$$

For moderate values of y, this results in a power law spectrum $F_\nu \propto \nu^{-\alpha}$ with spectral index $\alpha = \sqrt{9/4 + 1/y} - 3/2$. At frequencies $h\nu \gg kT_e$, the spectrum has a Wien tail. When $y \gg 1$, the entire spectrum has a Wien shape.

Observations, however, usually yield power law spectra with spectral index $\alpha \lesssim 1$ and a cut-off indicating a high temperature $kT_e \gtrsim 30 - 50$ keV. In this case, Eq. (II.1) and the above relationship between α and y would give a moderate $\tau_0 \simeq 1$, which violates the diffusion approximation. Under these circumstances, Eq. (II.1) is inadequate to estimate quantitatively the effects of Compton scattering, and the relationship between α, τ_0 and T_e is no longer valid. In order to treat the Comptonization process in this regime, one has to solve the full radiative transfer equation with the relativistic correction taken into account. It was found that a novel parameter replacing y can be determined by the plasma temperature T_e and the first eigenvalue of the kinetic equation for the problem. The general solution of the radiative transfer problem in isothermal plasma clouds was given by T94.

The problem of the spectral and angular distribution of photons undergoing scatterings in a hot plasma cloud, where the hard photon spectrum is formed by the photons escaping from the cloud after being scattered many more times than the average number of scatterings $\bar{u}$, has been analyzed by ST80 and ST85. The solution of this problem is closely connected to the distribution law of the number of scatterings $u = c\sigma_T n_e t$, where n_e is electron density; t is time duration of a photon random walking; σ_T is the Thomson cross section and c is the velocity of light . The exponential tail of the distribution over the scattering number is a typical feature of photons escaping from a limited region of space. The probability that a photon undergoes $u \gg \bar{u}$ scatterings is given by the asymptotic relation:

$$P(u) = A(\bar{u}, \tau_0) \exp(-\beta u), \qquad (II.2)$$

where the normalization constant $A(\bar{u}, \tau_0)$ depends on the distribution of source photons inside the plasma cloud /1-2/. The higher-order approximation forms for the parameter β were derived (/1/) for disk and sphere geometries with small as well as large optical depths:

$$\beta = \begin{cases} \dfrac{\pi^2}{12(\tau_0 + 2/3)^2}\left(1 - e^{-1.35\tau_0}\right) + 0.45e^{-3.7\tau_0}\ln\dfrac{10}{3\tau_0}, & \text{for disks} \\[3mm] \dfrac{\pi^2}{3(\tau_0 + 2/3)^2}\left(1 - e^{-0.7\tau_0}\right) + e^{-1.4\tau_0}\ln\dfrac{4}{3\tau_0}. & \text{for spheres} \end{cases} \qquad (II.3)$$

i) The optically thick case $\tau_0 > 1$

The upscattering Comptonization problem for plasma clouds with plasma optical depth $\tau_0 > 1$ in the wide range of plasma temperature reduces to solving the stationary equation for occupation number $N_1(x)$ (/1/):

$$x^2 N_1'' + x(x(1 + \delta) + 4)N_1' + (x^2\delta + 4x - \gamma)N_1 = -\psi(x), \qquad (II.4)$$

where $x = z/\Theta$;

$z = h\nu/m_e c^2$ is the dimensionless photon energy;
$\Theta = kT_e/m_e c^2$ is the dimensionless plasma temperature;
$\psi(x)$ is the initial spectrum of source photons;
$\mu = [1 + f_0(\Theta)]/(1 + 4.6z + 1.1z^2)$;
$\delta(x) = \frac{\mu'}{\mu} = -[(4.6 + 2.2x\Theta)\Theta]/[1 + 4.6x\Theta + 1.1(x\Theta)^2]$;
$f_0(\Theta) = 2.5\Theta + 1.875\Theta^2(1 - \Theta)$;
$\gamma(x) = \gamma_0(1 + 4.6z + 1.1z^2)[1 + 2.8(1 - 1.1\Theta)z - 0.44z^2]$;

and

$$\gamma_0 = \frac{\beta K_2(1/\Theta)}{\Theta K_3(1/\Theta)} \overset{\Theta \ll 1}{\simeq} \frac{\beta}{\Theta[1 + f_0(\Theta)]}, \qquad (II.5)$$

where $K_{2,3}$ are the modified Bessel functions of 2nd and 3rd order respectively. In order to find the analytical solution of Eq.(II.4), we should make a few simplifications in the equation coefficients δ and γ: i. Since δ depends weakly on energy x and it mainly influences the solution in the vicinity $x \sim 1$, one can replace the energy dependence of δ by the constant value $\delta = \delta(1)$. ii. We neglect the third and the fourth order terms of z ($z = \Theta x$) in the γ−polynomial with respect to the quadratic polynomial of z. With these assumptions the solutions to Eq. (II.4) can be expressed by Whittaker functions. In the case of low-frequency sources, such as $\psi(x) \propto \delta(x - x_0)$ with $x_0 \ll 1$, the spectrum of photons emerging from the plasma cloud is described by the simple formula

$$F_\nu(x, x_0) = \frac{\alpha_0(\alpha_0 + 3)e^{-sx}}{\Gamma(2\alpha_0 + 4)x_0} \left(\frac{x}{x_0}\right)^{-\alpha_0} \int_0^\infty t^{\alpha_0 + \epsilon - 1}(\rho x + t)^{\alpha_0 - \epsilon + 3}e^{-t}dt,$$

$$\text{for} \quad x \geq x_0 \quad (\text{II.6})$$

where F_ν is energy flux; $\Gamma(z)$ is the gamma function; $\alpha_0 = \sqrt{9/4 + \gamma_0} - 3/2$, $s = (1 + \rho + \delta)/2$, $\rho = \sqrt{(1 - \delta)^2 + 4a_2}$, $\epsilon = [2(\rho - 1) + a_1 + 2\delta]/2$, $a_1 = 7.4\gamma_0\Theta(1 - 0.42\Theta)$ and $a_2 = 13.5\gamma_0\Theta^2(1 - 1.05\Theta)$. For given optical depth τ_0 the values of β are uniquely determined by Eq. (II.3).

In /1/, we study the asymptotic spectral behaviour for the low energies $h\nu \ll kT_e$, and for the high energies $h\nu \gg kT_e$. We show that the spectrum becomes a power law with spectral index α_0 at low energies, and follows the law $x^{3-p}e^{-x(1+q)}$ ($p, q > 0$) at high energies [see Eq.(42) there]. Both asymptotic forms are contained in Eq.(II.6) in the limits $x \ll 1$ and $x \gg 1$ respectively. *This one formula recovers all of T94 results for the optical thick case, $\tau_0 > 1$.*

ii) Saturated Comptonization case $\gamma_0 < 1$ and modified ST80 solution

The Compton scattering cross section drops substantially when the electron temperature and the photon energy go up. In other words, a cloud is more transparent to higher energy photons and less capable of changing their energies and directions. Consequently even in plasma clouds with fairly large τ_0 and $kT_e > 100$ keV, most of the photons do not reach the Wien barrier $3kT_e$ and the Wien hard tail $x^3\exp(-x)$ is more or less suppressed. Quantitatively this effect could be accounted for by introducing the effective optical depth τ_R for photons at the very hard tail ($x > 1$, or $E > kT_e$). The effective optical depth (the Rosseland mean) is obtained as a result of averaging of the Compton scattering opacity over energy with the blackbody weight function at the electron temperature. The temperature dependence of τ_R can be described by the following approximation /12/:

$$\tau_R = \frac{\tau_0}{(1 + kT_e/39.2 \text{ keV})^{0.86}}. \quad (\text{II.7})$$

For simplicity, we assume that the photons suffering many scatterings have a quasi-uniform spatial distribution. This is certainly valid in the upscattering case where photons undergoing multiple scatterings are distributed in accordance with the first eigenfunction of the appropriate radiative transfer operator /1/. The reduced optical depth can be taken into account via the reducing factor r which is the probability for a photon to be scattered. Because of the preferential Compton scattering in the forward direction if one wants to find the lowest value of the reducing factor r, he should use the model of the distributed sources over the plasma cloud when all sources radiate outwards, with respect to the plasma cloud center, i.e.

$$r = \begin{cases} 1 - \dfrac{1 - e^{-\tau_R}}{\tau_R}, & \text{for disks,} \\ 1 - \dfrac{3}{\tau_R}\left[1 - \dfrac{2}{\tau_R} + \dfrac{2}{\tau_R^2}\left(1 - e^{-\tau_R}\right)\right]. & \text{for spheres} \end{cases} \quad (\text{II.8})$$

With this reducing effect taken into account, the spectrum is given by the modification of ST80 formula

$$F_\nu(x, x_0) = \frac{\alpha_0(\alpha_0 + 3)e^{-x}}{\Gamma(2\alpha_0 + 4)x_0} \left(\frac{x}{x_0}\right)^{-\alpha_0} \int_0^\infty t^{\alpha_0 - 1}(r^{\frac{1}{\alpha_0 + 3}}x + t)^{\alpha_0 + 3}e^{-t}dt,$$

$$\text{for} \quad x \geq x_0 \quad (\text{II.9})$$

From this formula, it follows that due to reduction of the scattering cross section, the Wien hump is suppressed and the power-law part is extended to higher energies. The reducing factor r also appears in the very hard tail (at $x \gg 1$, or $h\nu \gg kT_e$).

iii) The moderate optical depth case $\tau_0 \lesssim 1$

As was emphasized in ST85, T94 (see also discussion in Introduction of the present paper) and established by our Monte Carlo calculations for the cases of moderate optical depths $\tau_0 \lesssim 1$, photons which undergo $u \gg \bar{u}$ scatterings form a specific collimated radiation field concentrated along the maximum dimension of the plasma cloud (along the plane in disk geometry and the diameter in spherical geometry). As a result of averaging over the collimated angular distribution, the temperature dependent factor of the energy diffusion coefficient $\mu(z, \Theta)$ changes its form from that in Eq. (II.4) to $(8+19\Theta)/(8+15\Theta)$ $/1/$. Using the same idea of the photon escape at high energies, developed in the previous section the emergent spectrum can be described by Eq. (II.9) with γ_0 replaced by expression

$$\gamma_0 = \beta(8 + 15\Theta)/[\Theta(8 + 19\Theta)]. \qquad (II.10)$$

iv) Non-monochromatic source photon injection

$F_\nu(x, x_0)$ in Eqs. (II.6 and 9) are Green functions, that is, the upscattering spectra for monochromatic source photons $\psi(x) = \delta(x - x_0)$. One can derive the emergent spectra for arbitrary spectral distribution of source photons based on these Green functions. In particular, the upscattering spectrum I_ν for a Wien source injection, $\psi(x) = (ax)^3 \exp(-ax)/2$ $(a \gg 1)$, can be expressed in terms of F_ν as follows (see also $/2/$)

$$I_\nu = \frac{1}{2a}\left[F_\nu(x, 1/a)\Gamma(3 + \alpha_0, ax) + \frac{a\alpha_0}{(2\alpha_0 + 3)}(ax)^3 \exp(-ax)\right], \qquad (II.11)$$

where the number of injected source photons is normalized to unity and $\Gamma(3 + \alpha_0, ax)$ is the incomplete Gamma function.

III. COMPTONIZED SPECTRA

With the Monte Carlo method $/11/$, we calculated spectra of photons emergent from plasma clouds, disks, as well as spheres, with a wide range of temperatures T_e and optical depths τ_0. We compare these spectra with those obtained from the Generalized Comptonization models (T94) and from our new analytical results Eqs. (II.6, 9, 11). As was described in §II, there are two regimes of approximation employed in the analytical models. Regime 1 is for clouds with large optical depth ($\tau_0 > 1$), where Eq.(II.6) are applicable, while regime 2 corresponds to moderate optical depths ($\tau_0 \lesssim 1$). The approach developed in §II.iii brings together both approximations based on the modification of the ST80 solution, Eq. (II.9), by incorporating of the reducing factor r, Eq. (II.8). The ideas of introducing the energy dependent index $\alpha(x)$ in T94 (Eq.34 there) and the factor r in Eq. (II.9) are very similar: both of them take into account the extension of the power-law part at the expense of hard tail suppression because of relativistic effects. The transition from the optically thick case to the optically moderate one is regulated by changing the parameter γ_0 from that in Eq. (II.5) to that in Eq. (II.10).

In order to compare the spectra over a wider energy scale, we use $h\nu \times F_\nu$ instead of photon number $F_\nu/h\nu$, which decreases rapidly at high energies.

Figures 1 and 2 are examples of such comparisons. In Figure 1, we show the comparisons of Monte Carlo calculations (histograms) and analytical results (smooth curves) for photons emergent from disk clouds at $kT_e = 50$ (a), 100 (b) and with optical depth $\tau_0 = 1, 2, 5$ and 10. The curves are obtained from regime 1 approximations: the modified ST80 solution with the reducing factor r, Eq.(II.9) (solid), the new analytical solution of Eq. (II.6) [or the T94 analytical approximation Eqs.(35), (42)] (dash) and the model ST80 with correction of power-law index (dash-dot). It is seen

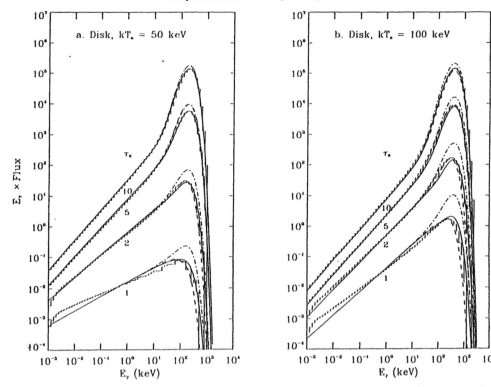

Fig. 1. The comparisons of the emergent photon spectra ($E_\gamma \times$ Flux vs E_γ) resulting from Comptonization of soft photon radiation for plasma disk at electron temperature $kT_e = 50$ keV with optical depths $\tau_0 = 1, 2, 5, 10$ or $\beta = 0.23, 0.11, 0.026, 0.0072$ respectively. The solid curves are the analytical approximations given by Eq. (II.9); dash lines present the exact subrelativistic analytical solution Eq. (II.6); dash-dot curves are obtained from model ST80 with power-law indices corrected by Eq. (II.5). Histograms are results of Monte Carlo calculations. The initial source photons have a blackbody spectrum at temperature T_r so that $\ln(T_e/T_r) = 10 - 20$. In Figures 1 and 2, the curves are re-scaled with the appropriate factors so as to separate curves with different optical depths. **Fig. 1b.** Same as in Figure 1a for plasma disk at $kT_e = 100$ keV.

that at low temperatures (Figs. 1a and b), all the Comptonized spectra from analytical models are in fairly good agreement with the Monte Carlo results except for τ_0 close to one. For large τ_0, the spectra have power-law portions with indices close to 1 (or 0 for F_ν). Among the three analytical models, the best agreement is achieved by the ST80 modified model with the reducing factor r, (Eq. II.9), which takes into account the suppression of the Wien tail due to relativistic effects. For diffusion regime ($\tau_0 > 1$) the subrelativistic analytical solution, Eq.(II.6) is also in good agreement with Monte Carlo calculation. It is worthwhile noting that this exact solution is the generalization of ST80 model for the case of subrelativistic photon energies and plasma temperatures.

In figure 2, we show the comparisons of Monte Carlo calculations (histograms) and analytical results (smooth curves) for photons emergent from spherical clouds at $kT_e = 100$ (a), 250 (b) and with optical depth $\tau_0 = 0.1, 0.2, 0.5$ and 1. The smooth curves are obtained from regime 2 approximations, that is, Eqs.(II.9 and 11) (solid) with γ_0 values given by Eq. (II.10). The dashed lines obtained by T94 analytical approximation [Eqs.(35), (44) there] practically coincides with the solid curves. As we have seen in Fig. 2b, regime 1 models are not very satisfactory for a disk with $kT_e = 100$ keV and $\tau_0 = 1$, or $\beta = 0.233$. Figure 2a shows that regime 2 models can give spectrum in fairly good agreement with the Monte Carlo results for a sphere with $kT_e = 100$ keV and $\tau_0 = 1$, which has a larger β value (0.67) than the disk in Fig. 2b. For smaller optical depths ($\tau < 1$), the agreement is still very good. The approximations are seen to be applicable to spheres of $\tau_0 = 0.1 - 1$.

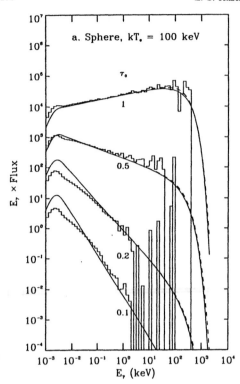
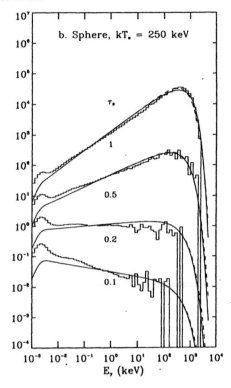

Fig. 2. The comparisons of spectra resulting from the Comptonization in plasma sphere with $kT_e = 100$ keV and $\tau = 0.1, 0.2, 0.5, 1$ or $\beta = 2.63, 2.01, 1.20, 0.67$ respectively. The curves are obtained by analytical Eqs. (II.9 and 11) with γ_0 given by Eq. (II.10); dash lines present the T94 analytical approximation [Eqs. (35 and 44) there] and they practically coincide with the solid lines. Histograms are results of Monte Carlo method. The initial source photons have a blackbody temperature $T = 0.511 \times 10^{-3}$ keV. **Fig. 2b** Same as in Figure 2a for plasma sphere at $kT_e = 250$ keV.

In the process of comparison, we also confirmed the T94 prediction that within certain range of temperature and optical depth and with source photon energies sufficiently low, *i.e.* when the Comptonization parameter $y \approx \Theta/\beta$ is of order unity /11/. the spectra emergent from plasma clouds depend on two parameters only – the plasma temperature T_e and parameter β given by Eq. (II.3). In other words, the photons from clouds with different geometry and low-frequency photon injection will have the same Comptonized spectra as long as they have the same temperature and β. This character of Comptonization allows us to apply the spectral knowledge obtained from a simple cloud geometry to cloud more complicated shapes . It also allows us to treat plasma clouds in a uniform way disregarding their geometries, such as in Figure 3, where we define an applicability region where Eqs. (II.6, 9 and 11) are insensitive to the conditions of source photons so that they can be applied to general situations.

The comparison result is summarized in Figure 3, where β and kT_e are used as two parameters characterizing the plasma clouds from which Comptonized radiation is emergent. The dashed lines at $kT_e = 5$ keV and 500 keV represent the lower and upper temperature limits of our comparison. Below 5 keV, upscattering is weak and electrons in the plasma can be considered as cold. Above 500 keV, other processes such as a pair production become important and should be taken into account in modeling the emergent radiation spectra, which are beyond the scope of this study. Between these temperatures, comprehensive comparisons were made for spectra from disk and sphere clouds with τ_0 ranging from 0.1 to 20. We established a diagonal region in $\beta - kT_e$ coordinates in which Eqs. (II.6, 9 and 11) agree well with the Monte Carlo results. This region can be considered as one where emergent spectra are insensitive to the source conditions and the analytical equations based on a special source distribution can be used for general situations. The figure also indicates

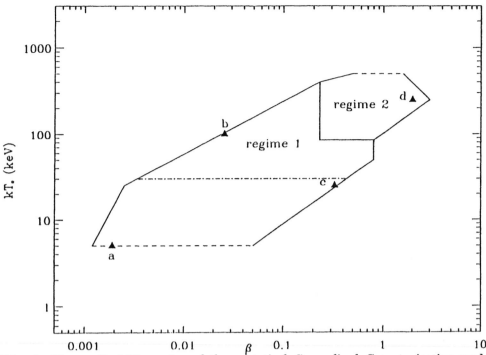

Fig. 3. The applicability region of the analytical Generalized Comptonization models in $\beta - kT_e$ parameter plane as established by comparisons with the Monte Carlo calculations. Within the region, the emergent spectra are insensitive to the source photon conditions so that the analytical models can be applied to general situation. Regimes 1 and 2 correspond to the two approximation models for optically thick and thin clouds respectively.

the applicability region where regime 1 and regime 2 apply. It is seen that, in general, regime 1 applies to plasma clouds having large optical depth (small β). Regime 2 is good for plasma clouds having rather small optical depth (large β) and Comptonized spectra are approximated by those photons which undergo repeated backward and forward scatterings (ST85, T94). The applicability parameter region is bounded to the right by a line

$$kT_e(\text{keV}) \approx 90\ \beta^{0.95}. \qquad (\text{III.1})$$

This line can be considered as the limit beyond which the analytical equations break down because of the upscattering effect inefficiency. Comptonization beyond this boundary is weak and the emergent spectra are small variations of the initial photon spectra. Above and to the left, the region is bounded by a line

$$kT_e(\text{keV}) \approx 1000\ \beta^{0.62}. \qquad (\text{III.2})$$

Beyond this boundary are high temperature and optically thick clouds for which analytical approximations of Eqs. (II.6, 9 and 11) do not match Monte Carlo results unless the source photon distribution is given by the first spatial distribution eigenfunction /11/. Thus this boundary represents the second limitation of the analytical results.

Originally, regime 2, the moderate optical depth approximation, was intended to be implemented in a region $kT_e \gtrsim 100$ keV and $\tau_0 \lesssim 1$ or $\beta \gtrsim 0.2$. But the temperature dependence of the energy diffusion is getting weaker with decreasing plasma temperature and this regime can also be applied to the region below ~ 30 keV (area below the dash-dotted line). In fact, in this region, the approximations of both regimes give rise to virtually identical spectra that fit well with the Monte Carlo results.

It is seen from Figure 3 that the applicability region of the analytical method covers a fairly large area, much larger than the previous model ST80 does, which is only the lower-left corner of our present region. The parameters derived from most of the observations so far fall in this area. For example, the best fit of analytical result to the observational data of Cyg X-1 obtained by EXOSAT, GRANAT and CGRO/OSSE /13,14/ gives $kT_e = 153$ keV and $\beta = 1.02$ /1/, which falls in the regime 2 region. Similarly, the regime 1 approximation was applied to fit the NGC 4151 data from CGRO/OSSE /15/ and Ginga /16/, which gave $kT_e \simeq 45$ keV and $\beta \simeq 0.2$ /1/. It is our plan to employ the analytical and Monte Carlo techniques described in this study to conduct a systematic analysis of the existing and forthcoming observational data.

IV. CONCLUSIONS

We introduced an exact solution of the Kompaneets equation for the diffusion regime ($\tau_0 > 1$) recovering T94 results and the ST80 solution modified with the reducing factor r. We also presented an efficient Monte Carlo technique for photon Comptonization in hot plasma clouds. The comparison of results from these two methods showed that the General Comptonization models are applicable to a wide variety of plasma clouds, covering a range of $\beta \simeq 10^{-3} - 2$ and $kT_e \simeq 5 - 500$ keV in the $\beta - kT_e$ parameter plane. We showed that within this parameter region 1) the emergent spectra are insensitive to the spectral and spatial distribution of source photons; 2) for the same temperature T_e and the same β, nearly identical upscattered spectral shapes are formed irrespective of the cloud geometries; and 3) the analytical models provide fairly good approximations for the emergent spectra, especially their power-law parts.

The new analytical technique provides effective tool for data analysis of observations from BHC and AGN sources obtained by such instruments as OSSE and BATSE.

Acknowledgements: L.T. wants to thank NASA and NRC for support under NRC Senior Research Associateship at the LHEA in GSFC. Also L.T. thanks Nick White for support and encouragement in all respects and Yurij Lyubarskij, Demos Kazanas, John Contopoulos, Natalie Mandzhavidze for discussion and assistance. XMH would like to thank R. Ramaty and R.E. Lingenfelter for their suggestions and valuable discussions in developing Monte Carlo codes for Comptonization in hot plasma.

REFERENCES

1. Titarchuk, L., Ap.J. **434**, 313 (1994).
2. Sunyaev, R.A. & Titarchuk, L., Astron. Astrophys. **86**, 121 (1980).
3. Pozdnyakov, L.A., Sobol, I.M., & Sunyaev, R.A., in Astrophysics and Space Physics Reviews, Soviet Scientific, ed. R.A. Sunyaev **2**, 189 (1983).
4. Zdziarski, A.A., Ap.J. **303**, 94 (1986).
5. Lightman, A.P., & Zdziarski, A.A., Ap.J. **319**, 643 (1987).
6. Prasad, M.K., Shestakov, A.I., Kershaw, D.S., & Zimmerman, G., J.Quant.Spectrosc.Rad.Transf. **40**, 29 (1988).
7. Nagirner, D.I., & & Poutanen, Yu. J., in Astrophys. Space Phys. Rev. (ed. R.A. Sunyaev)(N.Y.: Harwood Academic Publishers) **9**, 1 (1994).
8. Haardt, F., Ap.J. **413**, 680 (1993).
9. Haardt, F. and Maraschi, L., Ap.J. **413**, 507 (1993).
10. Sunyaev, R.A. & Titarchuk, L., Astron. Astrophys. **143**, 374 (1985).
11. Hua, X. & Titarchuk, L., Ap.J. in press , (1994).
12. Paczynski, B., Ap.J. **267**, 315 (1983).
13. Done, C., Mulchaey, J.S., Mushotzky, R.F., & Arnaud K.A., Ap.J. **395**, 275 (1992).
14. Grabelsky, D.A., et al., in Proc. of the First Compton Observatory Symposium, 1993 (AIP, St. Louis), p. 345.
15. Maisack, M., et al., Ap.J. **407**, L61 (1993).
16. Yaqoob, T., et al., MNRAS **262**, 435 (1993).

 Pergamon

Adv. Space Res. Vol. 19, No. 1, pp. (1)109–(1)112, 1997
Published by Elsevier Science Ltd on Behalf of COSPAR
Printed in Great Britain
0273–1177/97 $17.00 + 0.00

PII: S0273-1177(97)00044-6

PARTICLE ACCELERATION AND HIGH-ENERGY EMISSION IN THE EGRET AGNS

Evonne Marietta and Fulvio Melia

Steward Observatory, University of Arizona, Tucson AZ 85721, U.S.A.

ABSTRACT

Prior to the EGRET observations, jet formation models generally treated the acceleration and radiation mechanisms separately, since very little was known of the physical environment where the particles are initially energized. Because the high-energy emission from these sources presumably originates close to the central engine, the EGRET spectral measurements offer us the first opportunity to seriously model the early jet formation phase within $\sim 10 - 100$ Schwarzschild radii of the nucleus. A viable mechanism for producing the high-energy gamma-rays is the Compton upscattering of ambient low-energy photons (e.g., from a disk) by relativistically moving particles. However, it is well known that the resulting Compton drag on the particles can significantly retard their progress, which results in a particle-photon-induced resistivity. Thus, if the energizing force on the particles is associated with an AGN magnetospheric phenomenon, as many have hypothesized, the electromagnetic acceleration is fully dependent on the magnitude of the photon drag and hence is connected directly to the radiative emission itself. These two processes, i.e., the particle acceleration and the associated Compton upscattering, eventually need to be treated self-consistently. Our primary goal in this preliminary set of calculations is to establish the range of blackhole masses, the spin rate, and the permissible magnetic field configurations that produce spectra and fluxes consistent with the EGRET observations. This analysis also predicts the degree of beaming and inclination effects, both of which should bear directly on theories that attempt to unify the radio-loud AGNs into a single class. As an illustration, we apply our model to QSO CTA 102 and obtain a reasonable agreement with the observed flux and spectral index.

Published by Elsevier Science on behalf of COSPAR

INTRODUCTION

Many active galactic nuclei (AGNs) have been observed to have striking radio jets or strong radio core emission, whose origin is still not well understood. Because many quasars are also luminous X-ray sources whose strength is correlated with the radio core emission /1/, it is believed that understanding the high-energy emission, which probably originates close to a central, supermassive blackhole, will yield important insights into the formation of the large-scale jets. Until recently, however, the theoretical modeling of the jets has been hampered by the lack of hard X-ray/gamma-ray observations. EGRET on the Compton GRO, which was designed to measure photon energies from 20 MeV to 30 GeV, has now detected over 38 AGNs /2/. Most of them are quasars, at least five are BL Lac objects, and at least 10 are Optically Violent Variables. They are all radio loud, flat spectrum objects ranging from a z of 0.03 to more than 2.0. This gamma-ray emission is most likely associated with the particle acceleration region near the blackhole, thereby providing tantalizing clues into the physical processes acting in this enigmatic environment.

The retardation effects of Compton scattering on the particle dynamics has been recognized ever since it was discovered that high-velocity jets could be decelerated by particle-photon collisions

---

[1] NASA Graduate Student Researcher Fellow.
[2] Presidential Young Investigator.

/3/. This mechanism is sometimes used to account for the low terminal Lorentz factors ($\gamma_\infty \lesssim 10$) in superluminal sources even when the initial particle velocity is quite high /4/. Subsequently, Melia & Königl /5/ demonstrated that the Compton deceleration due to scatterings with accretion disk photons could transfer most of the jet's mechanical energy into an observable flux of high-energy radiation, accounting for both the low values of γ_∞ and the relatively flat X-ray spectra seen in some sources. A wide variety of particle energizing mechanisms have been proposed, ranging from radiation pressure-driven flows /3/ to electromagnetically-driven self-similiar jets /6/. The most enduring models are the electromagnetic ones because they alone appear to have the capability of accelerating particles to the relativistic speeds necessary to account for the radiative characteristics of superluminal sources. The preliminary spectral calculations reported here are based on the electromagnetic acceleration of charged particles within an approximately force-free blackhole magnetosphere, as first described by /7/. Following /5/, we assume the presence of a luminous accretion disk (often invoked to account for the "blue" bump observed in many AGN spectra) whose radiation scatters with the accelerating particles. The magnetosphere thus retains a small resistivity due to the particle-photon interactions that couple the acceleration of the particles and their concurrent radiative emission, but the fields remain close to their force-free intensities.

THE PHYSICAL MODEL

The structure of a force-free magnetosphere was first obtained by /7/ using the results of earlier work with pulsars /8/. MacDonald & Thorne /9/ recast the Blandford-Znajek theory using a universal-time absolute-space formalism. More recently, Okamoto /10/ refined the model by adding details on the structure of the currents and charge densities throughout the force-free region. In the work proposed here, the field structure is allowed to deviate *slightly* from the force-free condition due to the finite resistivity induced by the particle-photon interactions.

To check that the magnetospheric particles scattering with the accretion disk radiation can produce photons within the EGRET energy range, we can estimate the equilibrium lorentz factor γ_{eq} (at which the electromagnetic force is balanced by the Compton drag) assuming that the geometrically thin, optically thick disk /11/ radiates as a sum of blackbodies with a characteristic temperature $T \approx 10^5 K$. We find that even with such a high temperature increasing the photon drag and an assumed electric field only 10^{-6} times the strength of the magnetic field, a particle can be accelerated to $\gamma_{eq} \approx 10^4$. A particle with $\gamma_{eq} \approx 10^4$ can easily upscatter photons to energies $\gtrsim \gamma_{eq}^2 4kT \approx 3.5$ GeV, as observed by EGRET. With this value of γ_{eq} and T, the complicating effects due to particle cascades can still be avoided since the boosted disk photon energy in the particle rest frame is well below the pair creation threshold.

A reasonable jet model should also be able to produce a luminosity of order 10^{45} ergs/sec (assuming beaming). In this picture, the jet power is expected to be roughly $n_s n_\gamma \sigma_T c \gamma^2 \epsilon V$ where $n_\gamma \sigma_T c \gamma_{eq}^2 \epsilon$ is the average power emitted per magnetospheric electron, n_s is the average particle density, and V is the assumed total emitting volume. For typical values in the force-free magetosphere, the power is $\approx 10^{45} (\gamma_{eq}/10^3)^2 (B/10^4 \text{ G})(10^9 \ M_\odot/M)(a/a_{max}) \ (T/10^5 \text{ K})^4 (L/.01 \text{ pc})^3$ ergs/sec, where a is the rotation parameter. The preliminary indications are that a careful treatment of the particle dynamics and emissivity will produce the observed spectral characteristics of the EGRET AGNs.

An important assumption in this work is that the finite (though small) resistivity required to power the radiation field will not significantly alter the approximate force-free structure of the fields. This can be justified by comparing the intensity of the resulting non force-free magnetic field B_* to its force-free counterpart B_{ff}. B_* is given approximately by the condition $(J_* - \rho_e v) \approx \sigma(E_* + v/c \times B_*)$, where J_* is the current, ρ_e is the charge density moving at velocity v, and σ is the finite conductivity. Assuming that $J_* \gg \rho_e v$, $v \to c$, and $E_* \approx B_*$, we can take $B_* \approx J_*/\sigma$ as a ball-park estimate of the deviation from the force-free field. Using the particle-photon interaction rate to estimate the conductivity, we find that $B_*/B_{ff} \sim 10^{-5}/B_{ff} \ (T/10^5 \text{ K})^3$, so that the deviation from the force-free condition is less than 5% for $T \lesssim 10^7$ K when $B_{ff} \sim 10^4$ G the field intensity assumed by /12/. This temperature is well above the value $(2 - 3 \times 10^4$ K) inferred by

/13/ from the observation of a UV excess in the spectra of several AGNs and higher than the value 5×10^5 K inferred from the soft X-ray luminosity of PG 1211+143 /14/. Therefore, for reasonable values of the magnetic field and a disk temperature, the finite resistivity does not significantly alter the overall magnetospheric field structure.

To determine the location and structure of the high energy emitting region, we assume the analytic (uniform) magnetic field structure specified in Eqn. 3.5 of /15/. In this simple configuration, the uniform poloidal magnetic is frozen in and corotating with the accretion disk in the outer regions of the magnetosphere while in the inner regions it threads the event horizon of the central blackhole. In between the inner edge of the accretion disk and the event horizon the the uniform poloidal field passes freely through the equatorial plane. The force-free magnetosphere has a toroidal field which, in conjunction with the uniform poloidal field, results in particle trajectories which wind around the central object. The well-defined velocity field associated with these trajectories necessarily implies that only a specific region of the magnetosphere, namely the region within which the particle velocity is within $50/\gamma$ of the LOS, can contribute to the high-energy emission, and that the location of the emitting region changes with inclination angle. An illustrative image of this emission region, neglecting the effect of the disk on the toroidal field, for the uniform magnetic field is shown in Figure 1 for typical AGN parameters and an inclination angle of $0°$. Figure 1 shows the volume within which the particles do not contribute to the upscattered spectrum. For the uniform magnetic field, the contributing region is located along the axis and directly above the accretion disk where the toroidal field weakens. The location and size of the high-energy emitting volume clearly will have important implications for any unification theory that attempts to attribute some or all of the observed differences among classes of AGNs to an inclination effect.

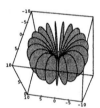

Fig. 1. High energy emitting region for $B = 10^4 G$, $M = 10^9 M_\odot$, $a = .5a_{max}$, and $\gamma = 10^4$.

The expected spectrum is a superposition of power-law components arising in different portions of the magnetosphere. The overall spectral index depends on the weighted distribution of these components, which in turn is a reflection of the magnetospheric structure and the actual temperature distribution in the disk.

DISCUSSION OF THE RESULTS

As noted earlier, our primary goal is to establish the range of physical parameters and the source geometry that can account for the observed high-energy fluxes and spectral indices. To this end, we simulate the particle acceleration and deceleration effects by the following analytic form for the Lorentz factor: $\gamma = 100 + 5.0/(e^z - e^{.0999})$.

This simplified, though restricted, particle behavior will be improved upon in the next generation of calculations, where the dynamics and the concomitant radiative characteristics will be determined self-consistently. The Compton scattered radiation is calculated according to the prescription derived by /16/. The spectrum is given as an integral over the magnetospheric particle distribution (fixed by the structure of the magnetosphere), and the incident photon intensity corresponding to a flattened relativistic disk with a temperature profile $T_e(r) = [3\dot{M}f(x)/8\pi\sigma M^2]^{1/4}$, where $f(x)$ is determined by the dissipation rate in the disk /17/. We show in Figure 2 the predicted photon spectrum (uncorrected for redshift) for a source distance of 1 Gpc, and the following system parameters: $M = 3 \times 10^9 \ M_\odot$, $T_{max} = 10^5 K$, $B = 10^4$ Gauss, and $a = .5a_{max}$.

E. Marietta and F. Melia

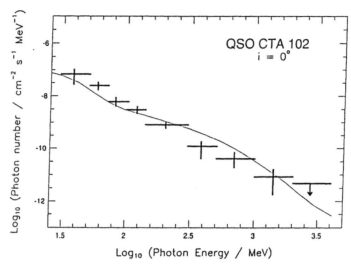

Fig. 2. Predicted photon spectrum. Observed data points for QSO CTA 102 are also shown.

For comparison, we also show the observed data points for QSO CTA 102, which at a redshift of $z = 1.037$, lies at roughly this distance for a cosmology with $q_o = .5$ and $H_o = 75$ km sec^{-1} Mpc^{-1}. The general agreement between the predicted and observed spectra is encouraging, particularly since the physical parameters are reasonably in the range permitted by independent observations and other theoretical arguments. The estimated luminosity of the majority of the EGRET objects is between 10^{45} - 10^{48} ergs/sec (isotropic emission) which in the Eddington limit implies a mass range of 10^8 - 10^{10} $M_\odot$. From equipartition arguments the characteristic magnetic field in the magnetosphere is expected to be on the order of 10^4 Gauss /18/. The maximum disk temperature was chosen in consideration of the observational evidence of the UV/soft X-ray excess observed in some AGN /14,15/. With a more complete survey of the permitted magnetic field configurations, a direct comparison between the predicted spectra, such as this, and the observations should provide us with a powerful technique for mapping the environment near the AGN central engines.

This work was supported in part by NSF grant PHY 88-57218 and NASA grant NAGW-2518.

REFERENCES

1. Browne, I. W. A. & Murphy, D. W. 1987, M.R.A.S. 226, 601
2. Hartmann, R.C. et al. 1993, The Second Compton Symposium, University of Maryland.
3. Abramowicz, M. A. & Piran, T. 1980, Ap.J. 241, L7
4. Phinney, E. S. 1982, M.N.R.A.S. 198, 1109
5. Melia, F. & Königl, A. 1989, Ap.J. 340, 162
6. Li, Z., Chiueh, T., and Begelman, M. C. 1992, Ap.J. 394, 459
7. Blandford, R. D. & Znajek, R. L., 1977, M.N.R.A.S 179, 433
8. Goldreich, P. & Julian, W. H. 1969, Ap.J. 157, 869
9. MacDonald, D. A. & Thorne, K. S., 1984, M.N.R.A.S 198, 345
10. Okamoto, I. 1992, M.N.R.A.S. 254, 192
11. Shakura,N.I. & Sunyaev, R.A. 1973,Astron. Astrophys. 24 , 337
12. Blandford, R. et al. 1990, Active Galactic Nuclei, Saa-Fee Advanced Course 20, pg. 169.
13. Malkan, M. A. & Sargent, W. L. W. 1982, Ap.J. 254, 22
14. Bechtold J., Czerny, B. ,Elvis, M., Fabbiano, G., & Green R. F., 1987, Ap.J. 314, 699
15. MacDonald, D. A. 1984, M.N.R.A.S 211, 313 255
16. Ho C. & Epstein, R.I. 1989, Ap.J. 343, 277
17. Page, D. N. & Thorne, K. S. 1974, Ap.J. 191, 499
18. Begelman, M. C., Blandford R.D., & Rees, M.J. 1984, Rev. Mod. Phys. 56, 255

Adv. Space Res. Vol. 19, No. 1, pp. (1)113–(1)116, 1997
© 1997. Published by Elsevier Ltd on behalf of COSPAR
Printed in Great Britain. All rights reserved
0273–1177/97 $17.00 + 0.00

PII: S0273-1177(97)00045-8

PAIR PRODUCTION NEAR ACCRETING BLACK HOLES

A. A. Belyanin* and R. F. van Oss**

* *Institute of Applied Physics, Russian Academy of Sciences, 46 Ulyanov St., 603600 Nizhny Novgorod, Russia*
** *Sterrekundig Instituut, Postbus 80 000, NL-3508 TA Utrecht, The Netherlands*

ABSTRACT

The non-relativistic temperature of the hard X-ray continuum in the spectrum of the black hole candidate 1E 1740.7-2942 during its hard state cannot be understood in terms of the conventional photon-photon pair production scenario's. The high pair creation rate during this state is indicated by the presence of a broad e^-e^+ annihilation line-like feature and needs the presence of a high temperature radiation field. We put forward a scenario in which the observed spectrum originates from the inner region of an accretion disk around a rapidly rotating black hole. For a not too high inclination angle, the gravitational redshift lowers the high local radiation temperature to the observed value. This mechanism necessitates a region of extremely strong gravity and is therefore unique for black holes. The presence of an annihilation line together with a low temperature X-ray continuum could therefore be considered as a black hole signature.

INTRODUCTION

Transient line-like features, which can be most easily attributed to electron-positron (e^-e^+) annihilation radiation, were recently detected from two hard X-ray sources that are known as galactic black hole candidates: 1E 1740.7-2942 /1/-/3/ and GRS 1124-68, also known as Nova Muscae 1991 /4/,/5/. The resemblance between the hard X-ray spectrum of 1E 1740.7-2942 (1E from now on) and the spectrum of Cygnus X-1 identified this source as a black hole candidate. The two main problems with the interpretation of the observation of the line is that firstly there is no apparent imprint of a hot pair production region on the spectra observed during the annihilation outbursts. A continuum extending beyond 0.5 MeV and indicating the presence of relativistic (thermal or non-thermal) plasmas, is absent. Therefore, it is not clear how to explain the high rate of pair production needed to provide the observed annihilation luminosity. Secondly, there is no correlation observed between the hard X-ray continuum and the appearance of the transient annihilation line. The spectra can be reproduced by simply imposing a Gaussian line profile on the X-ray spectra during their standard, persistent spectral states. We show that the first problem can be overcome if the observed continuum is interpreted as a gravitationally redshifted hot radiation component emitted by the inner disk region close to the event horizon of a rapidly spinning black hole. Locally, the radiation field produces enough pairs by photon-photon collisions to account for the line strength. The reason for taking the rotation of the black hole into account is that at increasing angular momentum of the black hole the radius of the last stable circular orbit (inner edge of the disk) moves closer to the event horizon. Consequently, the gravitational redshift of radiation emitted near the inner edge of the disk is maximized for a maximally rotating black hole (rotation factor $a = 0.998M$, /6/). However, in this case the inner disk rotates very rapidly and for a high inclination angle the redshift is counteracted by the Doppler blue shift.

We consider the following model for the hard state of 1E. The observed hard X-ray continuum is produced by an equatorial ring located at the radius of maximal energy release for a stationary disk. The transfer of radiation from this site to infinity is influenced by gravitational redshift, Doppler shifts, gravitational focusing and electron scattering in the pair wind. The first three effects are taken into account by applying a Monte Carlo simulation in which the polar angle at infinity of photons emitted from the disk are calculated. The direction of the photon in the frame comoving with the accretion flow is chooses randomly. We neglect the influence of electron scattering on the formation of the continuum. The radiation field from the equatorial ring produces pairs and blows them away by the radiation pressure because the local luminosity is super-Eddington for pairs. The pairs are driven by the radiation field, which is focussed towards the equatorial plane, onto the disk at somewhat larger radius. The positrons in the wind annihilate in the disk leading to the observed annihilation line.

Two free parameters enter this model: the rotation factor of the black hole a with $0 < a < 0.998M$ and the inclination θ_0 of the rotation axis with respect to the line of sight. Firstly we relate the local radiation field to the observed continuum and obtain the local radiation temperature and intensity depending on inclination and rotation factor. Secondly we calculate the pair production rate around the disk, compare this to the value inferred from the observed line strength, and obtain conditions for the inclination angle and rotation factor.

THE PAIR PRODUCTION RATE

Considering the shape of the observed continuum we take for the local intensity a spectrum of the Wien-type. The local intensity is furthermore assumed to be isotropic (no limb darkening):

$$I(x_e) = \frac{A}{4\pi} x^3 e^{-x_e/t}, \tag{1}$$

with $x_e = \epsilon_e/mc^2$ and $t = kT_e/mc^2$: the emitted photon energy and electron temperature in units of the rest energy of the electron. The factor A is a normalization constant fixed by the observed luminosity. We have numerically calculated the dependence of t/t_{obs} and A on inclination angle and rotation factor /7/. The observed temperature $T_{obs} \simeq 27$ keV, with $t_{obs} = kT_{obs}/mc^2$, is taken from spectral fitting of the continuum /1/. The value of A relates to the observed luminosity. We find that the observed temperature exceeds the local temperature for inclination angles higher than 60^0, more or less independent of a. The radiation temperature is red shifted maximally for a maximally rotating black hole at vanishing inclination. The dependence of A for fixed observed luminosity on inclination and rotation factor is weak, except when the black hole rotates maximally /7/.

The pair production rate for an isotropic radiation field is given by:

$$q = \int_0^\infty \int_0^\infty n(x_1)n(x_2)R(x)dx_1dx_2 \quad cm^{-3}s^{-1}, \tag{2}$$

with $n(x) = \frac{4\pi}{c}\frac{I(x)}{xmc^2}$ the photon number density and $R(x)$ the pair creation rate. We use the approximation for the pair creation rate from Coppi & Blandford /8/: $R(x) \simeq kc\sigma_T \ln x(x^2 - 1)/x^3$, with: $x > 1$ and $k = 0.652$. By using the assumed Wien spectrum we have:

$$q = (\frac{A}{mc^2})^2 \frac{kc\sigma_T}{c} \kappa(t), \quad \kappa(t) = 2 \int_1^\infty K_0(\frac{2}{t}\sqrt{x}) \frac{x^2 - 1}{x} \ln x dx, \tag{3}$$

with $K_0(z)$ the modified Bessel function of order zero. The pair production rate is a very steep function on temperature t through the function $\kappa(t)$. A doubling of the value for t, from $t = 0.05$ to 0.1 increases $\kappa(t)$ ten orders of magnitude. This implies that the radiation temperature is very well determined even if the pair production rate is known very poorly.

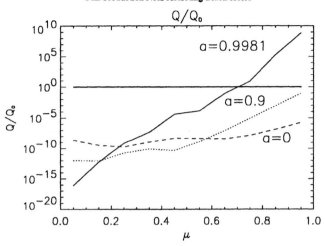

Figure 1: The total pair production rate in units of its lower limit Q/Q_0 as a function of the cosine of the polar angle of the observer for three values of the rotation factor a of the black hole. The lower limit on the pair production rate, $Q_0 = 2 \times 10^{43}$ s^{-1}, is fixed by the observed annihilation line strength. In case the rotation factor is lower than 0.9 the pair production rate calculated from the observed radiation temperature is below the rate implied by the observation. For the maximally rotating black hole the inclination angle is restricted to $\cos\theta_0 > 0.7$, or $\theta_0 < 46^0$.

Assuming the radial extend of the ring $\delta r = 0.1 r_g$ and the vertical extend of the pair production region equal to r_g we arrive at an expression for the total pair production rate Q /7/. We assumed a ten solar mass black hole and a luminosity of 10^{37} ergs s^{-1}. In Fig. 1 the dependence of Q/Q_0 on the cosine of the inclination angle for three values of the rotation factor of the black hole is shown. The value of $Q_0 = 2 \times 10^{43}$ pairs s^{-1} is the minimum value for the pair production rate obtained from the observed line strength. To explain the observed pair production rate a maximally black hole is needed and the inclination angle is restricted: $\cos\theta_0 > 0.7$, or $\theta_0 < 46^0$.

DISCUSSION

We have shown that a rapidly rotating black hole can be the powerful source of e^-e^+ pairs even if the observed radiation temperature of the continuum is rather small. The remaining problems are (i) the absence of correlation of annihilation outbursts with the continuum, (ii) the day-long variability of the hard states. The timescale of the order of 1 day is many orders of magnitude larger than all characteristic timescales in the inner disk. It is therefore unlikely that any local processes, e.g. instabilities in the inner disk, can account for the appearance of the hard state.

To explain these peculiarities, we suggest the following possibility. The hard X-ray luminosity of the innermost disk is evidently super-Eddington for pairs, even with the general-relativistic enhancement of the gravitational pull taken into account. Therefore, the created pairs are continuously driven away from the pair production region, forming the pair wind. In the normal state of 1E, most pairs in the wind fly away to a large distance from the source before slowing down and annihilating in the interstellar medium /9/ or within the magnetosphere supported by the accretion disk /10/. It was shown /11/,/10/ that both scenario's are consistent with the radio observations of 1E /12/, and can contribute to the formation of the narrow , unshifted annihilation line observed from the Galactic Center region.

We may expect that the angular distribution of the pair wind approximately follows that of the continuum radiation field emitted by the inner disk. The X-rays, isotropically emitted in the frame comoving with the disk flow, are strongly bent toward the equatorial plane while propagating outwards. Therefore, far from the inner edge of the disk the pair outflow is concentrated near the surface of the disk. This means that even a small change in the geometry of the accretion disk can strongly affect the fraction of pairs which is intercepted by the disk. Namely, if the polar angle, subtended by the accretion disk, occasionally increases, this would greatly increase the number of pairs hitting the disk, thus triggering the annihilation outburst.

The model explains in a natural way the absence of correlations between the annihilation outbursts and the continuum. The geometry of accretion flow can be changed due to the changing boundary conditions at infinity. This is expected if the source accretes directly from an inhomogeneous molecular cloud. In this case the day-long timescale of the hard state can be also understood.

REFERENCES

1. Bouchet L. et al., *Astroph. J.* 383, L45 (1991).

2. Sunyaev, R. et al., *Astroph. J.* 383, L49 (1991).

3. Cordier, R. et al., *Astron. Astroph.* 275, L1 (1993).

4. Sunyaev, R. et al., *Astroph. J.* 389, L75 (1992).

5. Goldwurm A. et al., *Astroph. J.* 389, L79 (1992).

6. Thorne, K.S., *Astroph. J.* 191, 507 (1974).

7. Van Oss, R.F., Belyanin, A.A., in preparation

8. Coppi, P.S., Blandford, R.D., *Mon.Not.R.Astron.Soc.* 245, 453 (1990).

9. Ramaty, R. et al., *Astroph. J* 392, L63 (1992).

10. Zheleznyakov, V., Belyanin, A., *Astron. Astroph.* 287, 782 (1994).

11. Misra, R., Melia, F., *Astroph. J* 419, L25 (1993).

12. Mirabel, I.F., Rodríguez, L.F., Cordier, B., Paul, J., Lebrun, F., *Nature* 358, 215 (1992)

Pergamon

Adv. Space Res. Vol. 19, No. 1, pp. (1)117–(1)120, 1997
© 1997. Published by Elsevier Science Ltd on behalf of COSPAR
Printed in Great Britain. All rights reserved
0273–1177/97 $17.00 + 0.00

PII: S0273-1177(97)00046-X

STUDY OF NONLINEAR ALFVEN WAVES IN AN ELECTRON–POSITRON PLASMA WITH A 3-D EM PARTICLE CODE

Ken-Ichi Nishikawa*, Jie Zhao**, Jun-ichi Sakai** and Torsten Neubert***

* *Department of Physics and Astronomy, The University of Iowa, Iowa City, IA 52242, U.S.A.*
** *Laboratory for Plasma Astrophysics and Fusion Sciences, Department of Electronics and Information, Faculty of Engineering, Toyama University, Toyama 930, Japan*
*** *Space Physics Research Laboratory, The University of Michigan, Ann Arbor, MI 48109, U.S.A.*

ABSTRACT

Results from 3-dimensional electromagnetic particle simulations of a magnetized electron-positron plasma with a relativistic electron beam ($\gamma = 2$) are presented. As part of the initial conditions, a poloidal magnetic field is specified consistent with the current carried by the beam electrons. The beam undergoes pinching oscillations due to the pressure imbalance between the inside and outside of the beam. A transverse two-stream instability is excited with large helical perturbations. In the process, background electrons and positrons are accelerated up to relativistic energy levels. Only background electrons are accelerated further along the z-direction due to the synergetic effects by both the excited transverse mode and the accompanying electrostatic waves caused by the breakdown of the helical perturbations.

Preliminary simulation results for a relativistic electron-positron jet show that shock waves excited at the jet head propagate into the jet and accelerate particles perpendicularly. The jet electrons are twisted at the front part of the jet.

INTRODUCTION

It is known that magnetized relativistic electron-positron plasmas exist in the region of the pulsar magnetosphere and in active galactic nuclei (AGN). The problem of strong electromagnetic waves in electron-positron plasmas is of prime importance in understanding the emission processes in pulsars and has been the subject of several investigations /1-5/. Such electromagnetic waves are capable of driving plasma particles to relativistic energies. This energetic particle distribution may account for the periodic radiation observed in association with pulsars /6-8/. It has been shown in many cases, that when the field amplitudes are large, the waves suffer modulational or filamentational instabilities /9,10/ and may become broken into highly localized pulses or filaments. However, these studies are all restricted to the theoretical analysis of wave propagation and linear dispersion relations. Since nonlinear effects in magnetized relativistic plasmas may play an important role in wave generation; interpretations considering only linear effects are not sufficient. Recently, relativistic electron-beam instabilities have been investigated for several applications. The "electron-hose" instability was proved to be excited by a relativistic electron beam in an unmagnetized plasma /11/. Intensive studies on the two-stream instability have been done in connection with its application to a high-power relativistic klystron amplifier (RKA) /12-14/.

We study the characteristics of a magnetized electron-positron plasma with a relativistic electron beam and a relativistic electron-positron jet by using a 3-D EM particle code /15-20/.

SIMULATION MODEL AND RESULTS

A. Excitation of waves by a relativistic electron beam

The code is a 3-D, fully electromagnetic and relativistic particle code /15/. The domain is a mesh of size $85\Delta \times 85\Delta \times 160\Delta$, where $\Delta(=1)$ is the grid size. Periodic boundary conditions are used for particles and fields in the z-direction, while radiating boundary conditions /21/ are used in the x- and y-directions. Particles which hit these boundaries are arrested there and then reflected with a thermal velocity randomly with some portion (50%). There are one million electron-positron pairs filling the entire domain uniformly and which keeps the domain charge neutral.

An electron beam in the z-direction is specified initially in a column with radius $r_{eb} = 4.47\Delta$ at the center of the xy-plane ($x_{cn} = 43\Delta$, $y_{cn} = 43\Delta$). Within the column, half of the electrons are beam electrons, while the other are background electrons. The temperature of the beam electrons is much colder than the temperature of the background electrons ($T_{eb} = 0.09T_e$). Initially only the beam electrons drift with $v_d = 0.865c$ (c is the light velocity) corresponding to a Lorentz factor $\gamma = (1-(v_d/c)^2)^{-1/2} = 2$. In addition to a homogeneous background magnetic field B_{z0}, a poloidal magnetic field B_θ consistent with the current density J_z carried by the beam electrons is included initially /21/ with $B_\theta/B_z = 0.48$. Other parameters are as follows: $m_p/m_e = 1, T_e/T_p = 1$ (T_p is the positron temperature), $\Omega_e/\omega_{pe} = 0.4$, $c/v_{tg} = 10.67$ (v_{tg} is the thermal velocity of the background electrons), $\omega_{pe}\Delta t = 0.1$, $\beta = 0.111$, $\lambda_{De} = 0.469\Delta$, $c/\omega_{pe} = 5.0\Delta$, and the gyroradii of electrons and positrons are $\rho_e = \rho_p = 1.17\Delta$.

Due to the pressure imbalance, the beam electrons are strongly pinched, therefore the slightly bunched beam electrons by the LTSI are a trigger of the transverse two-stream instability /22,23/. The gyroradii of the heated beam electrons become the same order of the radius of the current loop. Therefore, due to the sheared helical total magnetic field, the beam electrons are easily transversely shifted, which is an energy source of a transverse two-stream instability (TTSI). The TTSI excites helical perturbations which release the perpendicular electromagnetic energy showing that $B_\perp^2$ decreases after $\omega_{pe}t = 12$.

B. Simulation of a relativistic electron-positron jet

The large-scale properties and behavior of astrophysical plasmas have been investigated by Astro Fluid Dynamics (AFD) simulations in a wide variety of circumstances and astronomical context /25,26/. Simulations of jets by AFD simulations have bridged the gap between theory and observation. However, in order to understand the small-scale properties and dynamics of the pulsar wind, active galactic nuclei jets, radio galaxy jets, and protostellar jets, it is necessary to use a particle code. We use a fully relativistic, electromagnetic particle code to simulate wave-particle interactions which are essential for understanding the wave-driven mechanism in jets.

We have simulated a relativistic electron-positron jet ($\gamma = 1/(1-(V_{jet}/c)^2)^{1/2} = 5$). A first test exploring the jet system was run using a modest 85 by 85 by 160 grid and 1,000,000 electron-ion pairs. Initially, the ambient particles (1.0 electron-ion pair per cell) fill the entire box uniformly and a jet with a velocity $V_{jet} = 9.798c_{jet}^e$ (relativistic) is injected into the $+z$ direction, where c_{jet}^e is the jet electron thermal velocity. The jet particles are randomly injected around $z_0 = 25\Delta$ along the ambient magnetic field. The center of the initial jet is located at $x = y = 43\Delta$ with the radius 4Δ. The ambient electron thermal velocity is $v_{amb}^e = v_{amb}^p = 0.09375c$. The jet thermal velocity is $c_{jet}^e = c_{jet}^p = 0.1v_{amb}^e = 0.1v_{amb}^p$. The ambient magnetic field is chosen so that $\Omega_e/\omega_{pe}^{amb} = 0.4$ for the ambient electrons. The gyroradii of ambient electrons and positron are $\rho_e = \rho_p = 1.17\Delta$. Other parameters are as follows: $\beta_{amb} = 0.11$, $\beta_{jet} = 1.32$, $\lambda_D^{amb} = 0.469\Delta$, $\lambda_{skin}^{amb} = c/\omega_{pe}^{amb} = 5.0\Delta$, $\lambda_{skin}^{jet} = c/\omega_{pe}^{jet} = 3.54\Delta$, $c/v_{amb}^e = 10.67$. Alfvén velocity in the ambient plasma is $v_A/c = \Omega_i/\omega_{pi} = 0.05$; other parameters: $\eta = \rho_{jet}/\rho_{amb} = 1$, $M = V_{jet}/c_{jet}^e = 104.5$, and $K = P_{jet}/P_{amb}$.

Figure 1 shows the jet electrons in the simulation system at time step 200. The jet is twisted and locally pinched. This phenomenon seems to be caused by shock waves excited at the jet head. The shock waves propagate into the jet and accelerate particles perpendicularly, at this time the jet electrons are twisted as shown in Fig. 1.

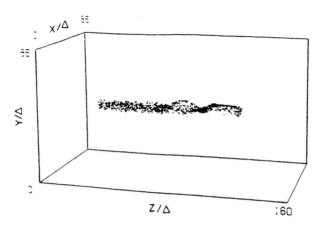

Fig. 1. The total jet electrons are plotted in the 3-D system at time step 200.

SUMMARY AND DISCUSSION

We have presented simulation results of a magnetized electron-positron plasma with a relativistic electron beam and a relativistic electron-positron jet using a 3-D electromagnetic particle code/24/. The dynamic behavior of a relativistic electron beam is classified into three phases: (1) Unlike our previous work /22/, the longitudinal two-stream instability is not dominant in the simulation results. The beam undergoes a pinching oscillation due to a pressure imbalance caused by the strong poloidal magnetic field generated initially by the current. This pinching effect triggers growth of a transverse two-stream instability. (2) A transverse two-stream instability is excited with the helical perturbations because the pinching makes the beam electrons do a large amplitude betatron orbit which spans the beam radius. The mode number of TTSI along the z-direction seems to be set by the initial LTSI. (3) The background electrons and positrons are heated up to relativistic energies associated with the excited transverse mode. We suggest that such high-energy electrons and positrons react with each other to generate γ-ray emission in a pulsar atmosphere. The beam electrons are also heated, though not to the same extent as the background electrons. The transversely heated electrons caused by the pinching effect leads to bunching of the beam electrons in the z-direction. The background electrons are further accelerated along the z-direction by the damped transverse mode and the electrostatic waves produced by the bunched beam electrons.

As expected, the simulation results of a relativistic electron-positron jet show small-scale complex characteristics of jet-head. Due to the complete dynamics of particles included in our simulations, the self-consistent acceleration of particles takes place in the jets. The charge density, which is not included in the AFD, plays an important role in the dynamics of jet. A particle code is an excellent tool for small-scale properties and behaviors of jets; on the contrary, AFD codes are an excellent approximation for large-scale global dynamics of jets.

REFERENCES

1. J-I. Sakai and T. Kawata, Nonlinear Alfvén wave in an ultra-relativistic electron-positron plasma, *J. Phys. Soc. Jpn*, <u>49</u>, 753-758, (1980).
2. A. B. Mikhailoskii, O. G. Onishchenko, and A. I. Smolyakov, Theory of low-frequency electromagnetic solitons in a relativistic electron-positron plasma, *Sov. J. Plasma Phys.*, <u>11</u>, 215-219, (1985).

3. L. Stenflo, P. K. Shukla, and M. Y. Yu, Nonlinear propagation of electromagnetic waves in magnetized electron-positron plasmas, *Astro. and Space Sci.*, 117, 303-308, (1985).

4. M. Y. Yu, K. Shukla, and L. Stenflo, Alfvén vortices in a strongly magnetized electron-positron plasma, *Astrophys. J.*, 309, L63-L66, (1986).

5. R. Bharuthram, P. K. Shukla, and M. Y. Yu, Current gradient-driven linear and nonlinear electromagnetic waves in a magnetized electron-positron plasma, *Astro. and Space Sci.*, 135, 211-218, (1987).

6. C. E. Max, Steady-state solutions for relativistically strong electromagnetic waves in plasmas, *Phys. Fluids*, 16, 1277-1288, (1973).

7. C. F. Kennel, F. S. Fujimura, and R. Pellat, Pulsar magnetospheres, *Space Sci. Rev.*, 24, 407-436, (1979).

8. A.C.-L. Chian, On the self-consistent solutions of pulsar plasma waves, *Astron. Astrophys.*, 112, 391-393, (1982).

9. A.C.-L. Chian and C. F. Kennel, Self-modulational formation of pulsar microstructure, *Astro. and Space Sci.*, 97, 9-18, (1983).

10. P. K. Shukla, Self-modulation of pulsar radiation in strongly magnetized electron-positron plasmas, *Astro. and Space Sci.*, 114, 381-385, (1985).

11. D. H. Whittum, W. M. Sharp, S. S. Yu, M. Lampe, and G. Joyce, Electron-hose instability in the ion-forced regime, *Phys. Rev. Lett.*, 67, 991-994, (1991).

12. H. S. Uhm, G. S. Park, and C. M. Armstrong, A theory of cavity excitation by modulated electron beam in connection with application to a klystron amplifier, *Phys. Fluid B*, 5, 1349-1357, (1993).

13. C. Chen, P. Catravas, and G. Bekefi, *Appl. Phys. Lett.*, 62, 1579-1581, (1993).

14. H. S. Uhm and C. Chen, Nonlinear analysis of the two-stream instability for relativistic annular electron beams, *Phys. Fluid B*, 5, 4180-4190, (1993).

15. O. Buneman, TRISTAN: The 3-D Electromagnetic Particle Code, *Computer Space Plasma Physics, Simulation Techniques and Softwares*, edited by H. Matsumoto and Y. Omura (Terra Scientific, Tokyo), 67-84, (1994).

16. T. Neubert, R. H. Miller, O. Buneman, and K.-I. Nishikawa, The dynamics of low-β plasma cloud as simulated by a 3-dimensional electromagnetic particle code, *J. Geophys. Res.*, 97, 12,057-12,072, (1992).

17. O. Buneman, T. Neubert, and K.-I. Nishikawa, Interaction of solar wind and Earth's field as simulated by a 3-D E-M particle code, *IEEE Trans. Plasma Sci.*, PS-20, 810-816, (1992).

18. O. Buneman, T. Neubert, and K.-I. Nishikawa, Solar wind-magnetosphere interaction as simulated by a 3D EM particle code, *Micro and Meso Scale Phenomena in Space Plasmas*, AGU Monograph, in press, (1994).

19. K.-I. Nishikawa, J. I. Sakai, J. Zhao, T. Neubert, and O. Buneman, Coalescence of two current loops with a kink instability simulated by a 3-D EM particle code, *Ap. J.*, in press, 10, (1994).

20. J. I. Sakai, J. Zhao, and K.-I. Nishikawa, Loops heating by D.C. electric current and electromagnetic wave emissions simulated by a 3-D EM particle code, *Solar Phys.*, in press, (1994).

21. K.-I. Nishikawa, O. Buneman, and T. Neubert, New aspects of whistler waves driven by an electron beam as studied by a 3-D EM particle code, *Geophys. Res. Lett.*, 21, 1019-1022, (1994).

22. J. Zhao, K.-I. Nishikawa, J. I. Sakai, and T. Neubert, Study of nonlinear Alfvén waves in an electron-positron plasma with 3-D EM particle code, *Phys. Plasmas*, 1, 103-108, (1994).

23. H. S. Uhm, A theory of two-stream instability in two hollow relativistic electron beams, *Phys. Fluids B*, 5, 3388-3398, (1993).

24. J. Zhao, J. I. Sakai, K.-I. Nishikawa, and T. Neubert, Relativistic particle in an electron-positron plasma with a relativistic electron beam, *Phys. Plasmas*, in press, (1994).

25. M. L. Norman, D. A. Clarke, and J. M. Stone, Computational astrophysical fluid dynamics, *Comput. Phys.*, MAR/APR, 138-151, (1991).

26. J. O. Burns, M. L. Norman, and D. A. Clarke, Numerical models of extragalactic radio sources, *Science*, 253, 522-530, (1991).

FUTURE SPACE MISSIONS TO PRIMITIVE BODIES

Proceedings of the B1.5 Meeting of COSPAR Scientific Commission B which was held during the Thirtieth COSPAR Scientific Assembly, Hamburg, Germany, 11–21 July 1994

Edited by

T. MUKAI

Department of Earth and Planetary Science, Faculty of Science, Kobe University, Nada, 657 Kobe, Japan

and

G. SCHWEHM

European Space Agency, ESTEC, Postbus 299, NL 2200 AG Noordwijk, The Netherlands

Pergamon

Adv. Space Res. Vol. 19, No. 1, pp. (1)123–(1)126, 1997
© 1997 COSPAR
Printed in Great Britain. All rights reserved
0273–1177/97 $17.00 + 0.00

PII: S0273-1177(96)00126-3

PLANNED OBSERVATION OF PHOBOS/DEIMOS DUST RINGS BY PLANET-B

H. Ishimoto, H. Kimura, N. Nakagawa and T. Mukai

Department of Earth and Planetary Science, Faculty of Science, Kobe University, Nada, 657 Kobe, Japan

ABSTRACT

An imaging CCD camera and an impact dust detector carried by the Mars mission (PLANET-B of ISAS, Japan) are potential instruments to search for the Phobos/Deimos dust rings proposed by theoretical investigations. We have reexamined the structure of dust rings based on the numerical simulation for dynamical evolution of ejecta from satellites around Mars, taking into account the gravity forces by the Sun, Mars, and the parent satellite and the solar radiation pressure forces on the ejecta as well as the perturbations due to the oblateness of Mars. It is found that the particles ejected with the mass of nearly 10^{-7}g from Phobos and 10^{-8}g from Deimos have a relatively longer survival lifetime against the collisions with Mars and the parent satellite. Deimos ring has higher number density (10^{-7}m^{-3}) than Phobos ring (10^{-8}m^{-3}). During roughly 2 years of mission life of PLANET-B from October 1999, a few ten particles across the rings will be detected by the dust detector with an effective area of 0.01m^2 and a threshold mass of 10^{-10}g. The imaging camera, on the other hand, will have a chance to look for the rings of the optical depth of roughly 10^{-7} to 10^{-9} from a favorable forward scattering angle of about $17°$.

© 1997 COSPAR. All rights reserved

DUST RING MODEL

The existence of dust rings near the Phobos/Deimos orbits has been proposed based on the theoretical works (e.g. Kholshevnikov et al. 1993, Juhász et al. 1993, Ishimoto and Mukai 1994). The "Phobos event", i.e., the irregularly fluctuation of plasma and magnetic field crossing the Phobos orbit detected by the Phobos 2 spacecraft (Riedler et al. 1989) has indicated the possible existence of a ring near the satellite's orbit. One of the scientific objects planned by the Mars mission (PLANET-B of ISAS, Japan), which will be launched in summer of 1998, is to search for a ring structure by using an imaging CCD camera and an impact dust detector on board. In this short paper, the detectability of the dust rings by the PLANET-B mission will be discussed, on the basis of the ring model improved from our previous paper (Ishimoto and Mukai 1994). The numerical simulation in our previous paper (Ishimoto and Mukai 1994) for the dynamical evolution of the particles ejected from the surface of Phobos was performed by taking into account the gravity of the Sun, Mars and Phobos as well as the solar radiation pressure forces on them. We have improved the simulation including the perturbations due to the oblateness of Mars, and applied it for Deimos as for Phobos. Since the details of the simulation will be presented elsewhere, only the resulting structure of dust rings is summarized below (see also Figure 1).

Our simulation has provided a cumulative number density of ring particles as functions of a distance from Mars and the mass of ring particle (see Figure 1). The width of the ring structure increases as the mass of ring particle decreases due to a relative importance of the radiation pressure forces on the small ring particles. The ejecta with a large ratio of solar radiation force to the solar gravity (a β ratio) hits on the surface of Mars quickly due to a large variation of the orbital eccentricity; otherwise the ejecta with a relatively smaller β ratio corresponding to larger mass moves on the orbit close to that of parent satellite. As a result, it is excavated on collision with the parent satellite. These results suggest that the ring particles have a narrow mass distribution. For Phobos/Deimos dust rings, the particles with masses of nearly $10^{-7}/10^{-8}$g, respectively, have a relatively longer survival lifetime around Mars.

PLANET-B MISSION

The PLANET-B mission will carry one dimensional CCD camera (Mars Imaging Camera (MIC))w
available 2560 pixels on one line and for three colors (RGB). MIC consists of a F1.4 lens with a fo
length of 30mm and an aperture of 25mm. The instrument will operate in line scan mode, making use of
image motion introduced by the spin of the spacecraft at 7.5 rpm. Its angular resolution is about 80 arcs
The details of the mission planning and the instruments on board of the PLANET-B mission will
summarized elsewhere. A final orbit of the spacecraft around Mars has not yet been fixed, although tw
cases shown later are under consideration. Then, we have studied the detectability of the proposed dw
rings for a polar orbit (an inclination angle of 75° to the ecliptic plane, case A in Figure 2) and an eclip
orbit (that of 148°, case B in Figure 2). Both have a periapsis at 150km above the surface of Mars and
apoapsis at a distance of 15 times of Mars radii from the center of Mars.

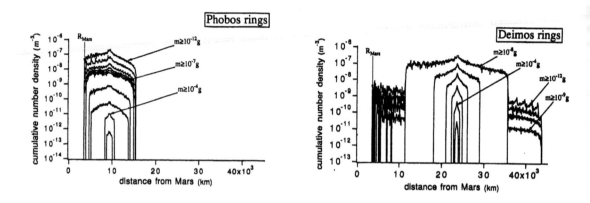

Fig.1. Ring structure for the particles with different masses m.

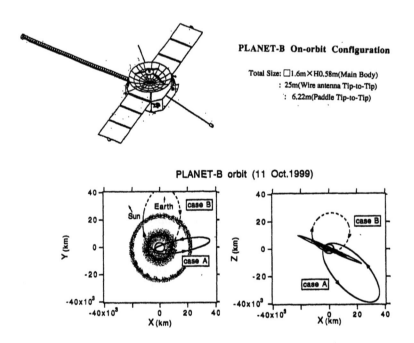

Fig.2. Two spacecraft's orbits planned now and Phobos/Deimos dust rings proposed
in the text. Case A denotes a polar orbit and case B is an ecliptic orbit.

MIC (Mars Imaging Camera) observation

Viking orbiter 1 tried to detect the dust rings and unknown satellites close to Mars, but could not (Duxbury and Ocampo 1988). One of the reasons is that the scattering angle at which Viking orbiter looked for the rings is large, i.e., about 145 to 160 degrees. It is well known that the forward scattering light becomes stronger roughly two or three orders of magnitude than the backward scattering light (see Figure 3). We are planning to observe the rings at small scattering angles during the mission life. The expected minimum scattering angle θ at which MIC can see is $17.3°$ for proposed Phobos dust rings and $17.7°$ for Deimos rings. The maximum optical depth τ along the line of sight for the dust ring during the mission becomes 10^{-9} for Phobos rings at $\theta=17.3°$ and 10^{-7} for Deimos rings at $\theta= 17.7°$. These values look smaller than that derived for the dust rings around Jupiter, i.e., about 10^{-5}, which were discovered while Voyager 2 was in Jupiter's shadow looking back toward the Sun (see Burns, Showalter and Morfill 1984). However, there still remain some uncertainties in the value of optical depth derived here. Namely, the ring particles are assumed to be single spheres consisting of silicate for simplicity. It is well known that the optical depth depends on the size distribution of particles, the optical constant of particle's material, shape and structure of the particle. Furthermore, the number density of ring particles has not yet been determined precisely as discussed below. Therefore, we cannot conclude now that the optical depth of the expected dust rings is too small to be seen.

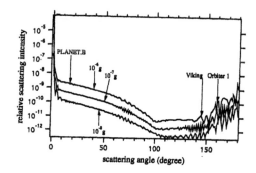

Fig.3. Phase function of ring particles. Mie theory was applied for each of ring particles with the complex refractive index of $1.48+i2.5\times10^{-5}$. The scattering angles observed by Viking Orbiter 1(min. and max.) and by planned PLANET-B are indicated by arrows.

MDC (Mars Dust Counter) detection

MDC is the advanced instrument from that used in HITEN (Igenbergs et al. 1991). It has the effective area of 0.01 m^2 and the minimum mass threshold ($>10^{-10}$g). Total number of impact particles is estimated from the dust ring model for two cases of the spacecraft orbit as a function of mass of particle (see Figure 4). In the case A (polar orbit), larger grains than 10^{-7}g for Phobos rings and 10^{-6}g for Deimos rings can not be detected because the spacecraft on the polar orbit never crosses the rings consisting of large particles. It should be noted that the values of total number of impact particles on the detector shown in Figure 4 have some uncertainties caused by the obscurity of the number density of ring particles. For example, we have neglected the secondary emission of particles by the collisions of ring particles on the satellite. Furthermore, the amount of ejecta depends on the structure of the surface of satellite, although we ignored such effect. An existence of regolith layer on the surface may produce more ejecta than those estimated here. Therefore we need further quantitative analysis for estimating the number density of ring particles.

CONCLUSIONS

In 1998, the Mars mission (PLANET-B of ISAS, Japan) will carry the instruments to observe the dust rings near the orbits of Phobos and Deimos satellites. Referring to our improved model for th Phobos/Deimos dust rings, we have found that the optical depth of dust rings becomes 10^{-9} for Phob rings and 10^{-7} for Deimos rings. In addition, the ring particles have a narrow mass distribution with 10^{-} for Phobos rings and 10^{-8}g for Deimos rings. From the analysis of the spacecraft orbit relative to th proposed dust rings, we have learned that there are occasions to detect the rings at the minimum scatteri angle of $17°$. Furthermore, the total number of impact particles on the detector will be about a few to particles during the mission life (about 2 years). These results sound that in our present dust ring model, is hard to find out the ring structure by using the instruments on board of the PLANET-B missio However, as discussed above, a future analysis of the ring structure will improve our knowledge about th parameters of the ring particles and consequently, it will enhance the number density of ring particles. W hope the dust rings around Mars appear during the Mars mission.

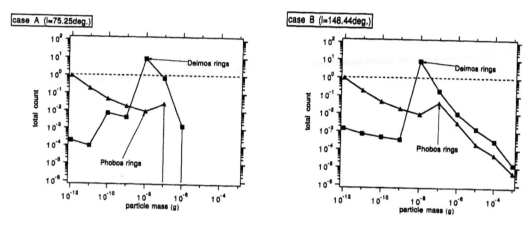

Fig.4. Total count of impact particles expected on MDC during the mission life of two years, as a function of particle's mass. Two cases are defined in Fig.2. Positive result will be obtained at least above the dashed line of one particle per 2 years.

REFERENCES

Burns, J. A., Showalter, M. R., and Morfill, G. E., The ethereal rings of Jupiter and Saturn, in:*Planetary Rings*, eds. R.Greenberg and A.Brahic, Univ. Arizona, Tucson , (1984)

Duxbury, T. C. and Ocampo, A. C., Mars:Satellite and ring search from Viking, *Icarus* 76, 160 (1988).

Igenbergs, E., Hüdepohl, A., Uesugi, K., Hayashi, T., Svedhem, H., Iglseder, H., Koller, G., Glasmachers, A., Grün, E., Schwehm, G., Mizutani, H., Yamamoto, T., Fujiwara, A., Ishii, N., Araki, H., Yamakoshi, K., and Nogami, K., The Munich dust counter:A cosmic dust experiment on board of the MUSES-A mission of Japan, in:*Origin and Evolution of Interplanetary Dust*, eds. A.C.Levasseur-Regourd and H.Hasegawa, Klumer Academic Publ., pp.45 (1991).

Ishimoto, H. and Mukai, T., Phobos dust rings, *Planetary and Space Sci.*42, p.691. (1994)

Juhász, A., Tátrallyay, M., Gévai,G., and Horányi, M., On the density of the dust halo around Mars, *JGR* 98, 1205 (1993).

Kholshevnikov, K. V., Krivov, A. V., Sokolov, L. L. and Titov, V. B., The dust torous around Phobos orbit, *Icarus* 105, 351 (1993).

Riedler, W., Möhlmann, D., Oraevsky, V. N., Schwingenschuh, K., Yeroshenko, Ye., Rustenbach, J., Aydogar, De., Berghofer, G., Lichtenegger, H., Delva, M., Svhelch, G., Pirsch, K., Fremuth, G., Steller, M., Arnold, H., Raditscht, T., Auster, U., Fornacon, K. -H., Schenk, H. J., Michaelis, H., Motschmann, U., Roatscht, T., Sauer, K., Schröter, R., Kurths, J., Lenners, D., Linthe, J., Kobzev, V., Styashkin, V., Achache, J., Slavin, J., Luhmann, J. G., and Russell, C. T., Magnetic fields near Mars: first results, *Nature* 341, 604 (1989).

 Pergamon

Adv. Space Res. Vol. 19, No. 1, pp. (1)127–(1)136, 1997
© 1997 COSPAR
Printed in Great Britain. All rights reserved
0273–1177/97 $17.00 + 0.00

PII: S0273-1177(96)00125-1

ROSETTA MISSION DESIGN

Martin Hechler

European Space Operations Centre, Robert-Bosch-Str. 5, 64293 Darmstadt, Germany

ABSTRACT

The prime objective of the ROSETTA Comet Rendezvous Mission is in situ analysis of cometary matter. Launched by ARIANE 5 in July 2003, the ROSETTA spacecraft will reach comet Wirtanen in 2011 utilising one gravity assists at Mars and two gravity assists at Earth and also performing fast flybys at the asteroids Shipka and Mimistrobell. It will enter into orbits around the comet and observe the nucleus and its environment from the distance of a few tens of km through perihelion passage in 2013. During its stay with the comet two scientific packages will be released by the spacecraft and will impact on the cometary surface at selected landing sites to augment the remote observations of the nucleus by in situ surface measurements. Near the comet, images of landmarks on the surface of the nucleus will be processed on the ground to derive spacecraft position and comet rotational state knowledge. Based on this knowledge safe orbit and surface package delivery strategies will be implemented. This paper gives a summary of the ROSETTA mission design as reflected in the Science Team Report /1/ with some extensions on recent developments, mainly in the area of mission analysis.
©1997 COSPAR. All rights reserved

INTRODUCTION

Background

Following the definition of the 'Planetary Cornerstone' in the long term plan of Scientific Projects of ESA in 1984 (Horizon 2000 /3/), ROSETTA had originally been conceived as a Comet Nucleus Sample Return Mission. To this objective the mission concept had evolved from 1984 to 1991 through a series of scientific and engineering efforts, since 1988 as a joint ESA/NASA project. Sample return mission and spacecraft concepts, based on a Mariner Mark II carrier spacecraft, had reached a considerable level of sophistication during a System Definition Study (1989 to 1990). 'Financial and programmatic difficulties experienced by NASA, the envisaged partner for the original mission concept, made it necessary to study alternative concepts in early 1992' /4/. Recent evolution in space experiment capabilities opened the possibility of 'taking the laboratory to the comet' rather than bringing a sample back to Earth, and still coming close to the original scientific objective of the planetary cornerstone as stated in /3/. In parallel to the discussions in the scientific community on a core European ROSETTA mission, engineering studies proved the feasibility of a comet rendezvous, with a spacecraft design which uses solar arrays for power generation up to 5.2 AU, and is launched by ARIANE. The mission included an experiment package to be dropped by the spacecraft onto the surface of the comet. In November 1993, this concept of ROSETTA has been selected as 'Cornerstone 3' of the ESA Scientific Program. At the time of project selection the target comet was Schwassmann-Wachmann 3, with a mission of shortest total duration, but on the limit of the mass budget /1/. In the meantime the reference target has been changed (Wirtanen) to gain sufficient mass margin for a second surface package. Studies on the system level and on key technologies are planned to refine and consolidate mission and spacecraft design of ROSETTA before the start of phase B of the project in 1997.

Mission Objective

Cometary nuclei and - to a lesser extent - asteroids represent the most primitive solar-system bodies. They are assumed to have kept a record of the physical and chemical processes that prevailed during the early stages of the evolution of the solar system. Analysis of comets and asteroids as a whole and of cometary material in particular, is expected to provide essential information on the provenance of meteorites and interplanetary dust and to improve our current understanding of the formation of the solar system /4/.

To this objective the ROSETTA mission will /4/:

- perform in-situ investigations of the chemical, mineralogical and isotopic composition, and the physical properties of volatiles and refractories in the nucleus,

- study the development of cometary activity, and its link with the characteristics of the nucleus (active areas; mantled areas),

- acquire complementary information on the diversity of asteroids from selected fly-bys.

Model Payload

The (straw-man) payload to be accommodated on the ROSETTA spacecraft consist of a set of 7 complementary instruments /4/:

- The *Remote Imaging System (RIS)* consists of a narrow angle camera (NAC 3.4° × 3.4° field of view) and a wide angle camera (WAC 17.2° × 17.2° field of view) both using a CCD array detector of 2000 × 2000 pixels, for remote sensing of the targets at visible wavelengths. The NAC will also be used for spacecraft optical navigation in conjunction with the star and target tracker.

- The *Visible and Infra Red Spectral and Thermal Mapper (VIRSTM)* uses an array detector (64 × 64 pixels) and 256 channels (from 3.5 μm to 5 μm) for determining the mineralogical composition of the targets and mapping the surface in terms of temperature.

- The *Neutral Mass and Ion Spectrometer (NGIMS)* is composed of one single focused and one double focused mass spectrometer for determination of the chemical, molecular and isotopic composition and density of gas and ions.

- The *Dust Mass Analyser (COMA)* is basically a mass spectrometer. It measures the mass of ionised molecules in a gas phase and provides elemental, isotopic and molecular composition.

- The *Scanning Electron Microprobe and Particle Analyser (SEMPA)* is using an electron microscope to give images of collected particles as small as 250 nm. Spectroscopy of the X-rays from the sample allows element detection and analysis for all elements above Na.

- The *Dust Production Rate and Velocity Analyser.*

- The *Plasma Instrumentation Package* to monitor the electron density and the solar wind flux.

The Surface Packages (45 kg each) have to be brought into contact with the cometary surface. Their straw-man payloads /1/ contain the following instruments (a core on both packages):

- The γ *Ray Spectrometer (GRS)* is used for determination of the major elements of the target composition.

- The α *Backscatter and X Ray Fluorescense Spectrometer (ABXRF)* is used for the determination of all major and most important minor elements of the target composition including Carbon, Nitrogen and Oxygen.

- The *Neutron Spectrometer (NS)* is used for determination of the hydrogen concentration by monitoring the neutron flux in two energy ranges: thermal and epithermal.

- The *In Situ Imaging System (ISIS)* consists of several monitoring and panoramic cameras used for in situ studies by characterising the local surface morphology to determine surface texture at varying resolutions (some mm per pixel).

- The *Thermal Analyser / Evolved Gas Analyser (TAEGA)* measures in a pyrolysis cell the thermal behaviour of cometary material and the chemical nature of the evolved gases.

- The *Permitivity Probe (PP)* measures electric properties of the material close to the surface.

- *Accelerometers* study the mechanical strength in the cometary crust during landing.

- *Sounding* of the interior structure of the comet (if feasible).

The Surface Science Packages will be designed as integrated independent, self-supporting units. They will be defined as PI instruments to provide for the most efficient integration of the different sensors.

MISSION OPPORTUNITIES

Reachable targets for a comet rendezvous are those short periodic comets with low inclination with respect to the ecliptic plane and perihelion radii near 1 AU. The aphelion radii of these objects are typically around 5.2 AU. For a dry spacecraft mass of the order of more than 1000 kg and an ARIANE 5 launch, multiple gravity assists at Earth, Venus and Mars are needed to reach the comets. Direct missions, or missions using one Earth gravity assist (ΔVEGA) are not yielding sufficient mass. Table 1 gives a catalogue of the

Table 1. ROSETTA Mission Opportunities 2003-2004

No	Rendezvous with Comet	Asteroid Flybys at	Type	Launch (date)	Arrival Perihelion (date)	Mission ΔV (km/s)	AR 5 Perf. (kg)	max. S/C dry mass (kg)
1	Wirtanen	Mimistrobell Shipka	MEE	2003/01/20	2011/08/27 2013/07/10	1.476	2990	1601
2	Schwassmann -Wachmann 3	Brita	ME	2003/07/18	2008/06/10 2011/10/12	1.924	2426	1106
3	Finlay	1990 OK	MEE	2003/07/17	2013/09/05 2014/12/09	1.837	2597	1224
5	Wirtanen	1982 DX3 1983 AD	VEE	2003/11/03	2012/10/23 2013/07/10	1.015	2698	1667
6	Haneda -Campos	Isis	VEE	2003/11/21	2013/06/23 2016/11/08	1.465	2571	1372
7	Schwassmann -Wachmann 3	1990 TJ	VE	2003/11/29	2008/10/28 2011/10/12	1.930	2352	1068
8	Finlay	1982 BB Lunacharsky	VEE	2004/05/11	2013/12/01 2014/12/09	1.073	2651	1614
9	Brooks 2	Carr 1983 WM	VEE	2004/05/25	2011/11/23 2014/05/23	1.740	2413	1170

best found comet rendezvous mission opportunities for an ARIANE 5 launch, using two or three gravity assists at Earth, Venus or Mars with a launch in 2003 and 2004. The stated ARIANE 5 launch mass is reduced by a 10% margin and an allocation of 50 kg for the launch window, assuming delayed ignition of the upper stage. The maximum possible spacecraft dry mass is calculated assuming a 75 kg launch adaptor, and incrementing the mission ΔV by 70 m/s for launch date variations, 160 m/s for cruise navigation, and 160 m/s for near comet orbit control. In addition 30 kg propellant for wheel offloading and attitude control are accounted for. The specific impulse of the spacecraft motor has been assumed to be 312 s. In the system definition study /2/ the required spacecraft dry mass was 1050 kg.

The asteroid flybys in the table are those which have been found to minimize the propellant requirements, in some cases only one flyby opportunity could be constructed. Missions to duToit-Hartley and Neujmin 2 were removed from the table for scientific reasons and, recently, to accommodate a second surface package, the set of mission opportunities was reduced to those three in bold print. For these missions the mass margin also gives more flexibility to the spacecraft design. Comet Wirtanen which has been chosen as target for the reference mission (Jan 2003 launch) and the first backup, also has been classified as the scientifically most attractive of the reachable targets. Studies have been started on the the ephemeris determination of comet Wirtanen based on the sparse set of observations usually collected for minor bodies, and taking into account non-gravitational effects in conjunction with gravitational perturbations by Jupiter. A well organised high quality astrometric campaign starting with aphelion observations in 1995 appear to be desirable.

MISSION SCENARIO

Reference Mission

Figure 1 shows the ecliptic projection of the reference mission interplanetary transfer to comet Wirtanen (first launch opportunity in the launch window on 2003/1/14). The main mission events are marked, the time ticks are every 100 days.

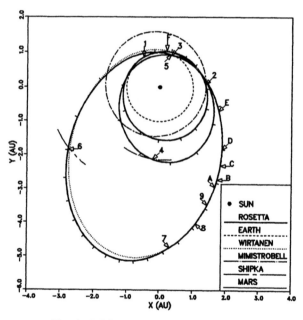

	event	day	date
1	Launch	0	2003/01/14
2	Mars flyby	955	2005/08/26
3	Earth flyby	1047	2005/11/26
4	Mimistrobell fb.	1340	2006/09/16
5	Earth flyby	1777	2007/11/27
6	Shipka flyby	2097	2008/10/12
7	Wirtanen rdv	3145	2011/08/25
8	Approach	3325	2012/02/21
9	Global mapping	3415	2012/05/21
A	Close observ.	3485	
B	Surf. pack. del.	3510	
C	3 AU from sun	3555	2012/10/08
D	Surf. pack. del.	3600	
E	2 AU from sun	3690	2013/02/20
F	Perihelion	3830	2013/07/10

Fig. 1. ROSETTA mission to Wirtanen

The arrival at the comet is at a sun distance of 4.78 AU. This may not allow to operate the on board cameras, because of the limited available power. Therefore the main orbit matching manoeuvre may be executed before on board detection of the comet. For the same reason actual near comet approach operations may be delayed until 4.2 AU sun distance is reached. Near comet operations (mapping) is planned not to start before a communications distance of less than 3.25 AU is reached, to have ground support from a second ESA (15 m) station. At larger distances the mission will be operated using the 30 m antenna at Weilheim (DLR) and, if available, the NASA/DSN.

Mission Phases

<u>Interplanetary Cruise and planetary gravity assists.</u> The ARIANE 5 launch from Kourou will first inject the

ROSETTA spacecraft into a 2 hours orbit around the Earth. The upper stage will be ignited with a 2 hours delay, before perigee passage, and will inject the spacecraft to the required hyperbolic escape conditions towards Mars. During the first few months after launch all spacecraft functions will be checked out, in particular those autonomous functions which the spacecraft needs to survive over long time intervals in the cruise phase without ground intervention and to reacquire ground control in failure cases. During this initial phase the orbit will be determined and correction manoeuvres will remove the launcher dispersion effect. After the checkout, it is planned to leave the spacecraft with a minimum of ground intervention for extended periods, regular ground operations activities will only be resumed 3 months before to 3 months after planetary and asteroid flybys, and before the arrival at the comet.

The Mars flyby after 955 days (almost two revolutions around the sun) will be at the lowest possible altitude (set to 200 km) and a manoeuvre of about 140 m/s will be performed near the flyby pericentre. This manoeuvre eventually can be suppressed at the cost of additional deep space manoeuvres and with a loss in the mission performance. Orbit correction manoeuvres required before and after the Mars flyby are estimated as 25 m/s in total (99-percentile). For the reference mission the targeting accuracy at Mars is influenced by the fact that the spacecraft will be seen at near zero declination from the Earth which will deteriorate the accuracy of conventional Earth based tracking. With range and Doppler from the 15 m stations only, the targeting error is estimated to be about 100 km (3σ). Therefore the flyby pericentre has been chosen 100 km above the Mars atmosphere. With additional tracking capabilities (e.g. the DSN or VLBI or differential tracking with a spacecraft in orbit around Mars) the targeting could be improved. Only 90 days after the Mars flyby the spacecraft will return to the Earth for a first time. The flyby altitude will be over 3000 km and the navigation will be uncritical because of the high precision of range and Doppler near the Earth. A manoeuvre of 60 m/s will be performed at the flyby. The second Earth flyby after another two years (4 years from launch) has a pericentre altitude of about 3000 km. In both cases orbit corrections of up to 10 m/s are foreseen.

Asteroid flybys. Between the two Earth gravity assists the spacecraft will pass through the asteroid main belt. At this occasion a first asteroid flyby with 3840 Mimistrobell was found to be most favourable in terms of propellant consumption. The relative velocity will be 6.1 km/s and the sun-asteroid-spacecraft angle during approach will be 58°. A second asteroid flyby can be accomplished with 2530 Shipka on the way out to the comet. The flyby velocity will be 13.2 km/s and the illumination angle 23°. Detection of the asteroid by the spacecraft narrow angle camera is expected about 8 days before the flyby. After detection the images can be used for optical navigation, orbit correction manoeuvres can be scheduled to improve the targeting accuracy to a level of about 20 km, from the initial ephemeris error of ground bases astrometry which is expected to be a few 100 km. The flyby point is chosen such that the asteroid to sun line is in the relative orbit plane. The nominal flyby distance has been chosen at least $\sim$ 500 km sunward, which is the limit of the turning rate for the viewing instruments mounted on the spacecraft body. The payload will be operated within 500 000 km distance from the asteroid. About one day after the flybys, a major orbit manoeuvre will re-target the spacecraft to the Earth or the comet respectively.

Comet approach. The orbit manoeuvre (1140 m/s) which puts the spacecraft into rendezvous conditions with the comet except for a residual drift rate ($\sim$ 100 m/s) will be performed before comet detection by the spacecraft camera, using ground astrometry knowledge of the comet ephemeris only. At this manoeuvre the sun distance will be 4.78 AU and the Earth distance will be 4.33 AU for the reference mission. The comet distance will be of the order of 5×10^7 km. The spacecraft will drift towards the comet until it reaches a sun and a communications distance which allows comet detection operations. The initial targeting of this drift will be to a point 100 000 km from the comet towards the sun. Figure 2 explains the approach strategy. Actual approach operations is planned to start at a distance of 500 000 km from the comet at which the high resolution camera can be expected to detect the nucleus.

The approach manoeuvre sequence will sequentially reduce the relative velocity to finally 10 m/s, within 90 days from the start of the approach. Each manoeuvre which reduces the relative velocity will also move the targeting point closer to the nucleus, in accordance with the improved relative orbit knowledge. At 1000 km distance the targeting bias will be 100 km. Image processing on the ground will derive spacecraft to comet directions in the inertial reference frame, using the fixed stars in the image background, which will improve the spacecraft state estimate relative to the comet, or better the comet ephemeris relative to the spacecraft.

A coarse estimation of comet size, shape and kinematics will be performed. The payload will be checked out. Below 50 cometary radii from the nucleus (mean nucleus radii of 1 to 5 km are expected) an estimate of the comet gravity constant and a refinement of shape and kinematics knowledge will be derived using optical observations of landmarks on the comet.

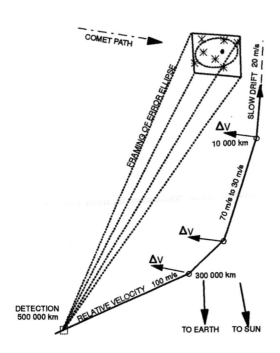 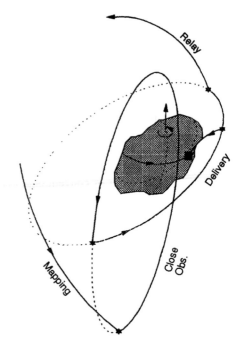

Fig. 2. Comet approach   Fig. 3. Near comet orbits

Near comet orbits (mapping and close observations). Figure 3 indicates the different phases from approach to surface package delivery. First, mapping strategies must be constructed to cover the nucleus surface globally at a prescribed resolution (2000 pixel across), and later to map selected areas with a high resolution, for the full range of possible gravitational and kinematic parameters (nucleus radius 1 - 5 km, density 0.2 -1.5 g/cm^3, rotation period 3 hours to ∞, free kinematic body motion - tumbling or regular spin). These comet properties will be largely unknown before arrival, only some knowledge on size and spin rate might be derived by sophisticated astronomical observations. This means the orbit strategies can not be fixed beforehand, they will be decided during the late approach phase. In spite of the low gravity (surface gravity 5×10^{-6} to 2×10^{-4} Earth-g) and the possibility of a very irregular shape of the nucleus (e.g. like Halley), rather stable spacecraft orbits around the nucleus are well possible. The sphere of influence of the comet, in which the sun attraction can be treated as a perturbation, will have a size from 7 to 20 nuclus radii (7 - 100 km) for above values of the mean radius and the density. The effects of the non-spherical shape of the gravity field will be of minor influence for distances above a few cometary radii. Therefore orbit strategies near the comet can be constructed using Kepler orbit approximations, they have then to be verified by simulations including all gravitational and non-gravitational (solar radiation pressure, comet outgasing) details.

Mapping will be done from eccentric orbits around the comet, with orbit parameters depending on comet gravity and spin properties. Usually the orbit plane will be chosen to contain the comet spin axis and the comet sun line. The period of the mapping orbits will be of the order of a few days. In many cases, only one half revolution will be necessary, starting near one pole, to produce a surface map by adjacent east-west coverage bands due to the comet rotation. Including the transfer to the mapping, and some time for navigation, in total 70 days have been allocated for the global characterisation. This will allow to achieve an almost complete coverage of the illuminated surface also in extreme cases of small, slowly rotating comets. From the mapping, comet shape and surface properties (physiography, roughness) will be determined, and a detailed kinematic and gravitational model using optical landmark observations will be derived. Areas on the surface will be selected for close observations. These close observations will then be done from a sequence of eccentric orbits which pass over up to five selected sites at altitudes below 1 nucleus radius. At least two orbit manoeuvres will be necessary per fly-over. All remote observations payload will be operated above the candidate sites (500 m × 500 m). 25 days have been allocated for close observations. Using the payload data, a decision will be taken to which site the first surface package will be delivered.

Near comet navigation. Important for the choice of the orbits will be the accuracy at which these orbits can be predicted, to guarantee spacecraft safety. The orbit determination near the comet will be primarily based on observations of natural landmarks on the surface by the wide angle target tracker. Together with the range and Doppler measurements from Earth, this will allow to simultaneously estimate 19 dynamical state variables, namely the spacecraft position and velocity relative to the comet (6 parameters), and comet position and velocity relative to the sun (6 parameters), the quaternions to define the axes of the comet nucleus relative to inertial (ecliptic) axes (4 parameters), and angular rates of the nucleus in body axes (3 parameters). In addition, for a typical case using five landmarks, there are 19 constants to be estimated, which are equivalent to the principal gravitational constant of the comet and the elements of the inertia matrix (7 constants related to the third order harmonics of the gravity field), and the positional coordinates of the landmarks (12 constants).

A variety of parametric simulations (/5/, /6/) have been performed to prove the feasibility of the near comet orbit strategies and the related landmark navigation process, for different phases of the near comet sequence, comparing theoretical standard deviations produced by covariance analysis with simulated errors, and for the whole range of assumptions on the dynamic and kinematic properties of the cometary nucleus, its size and shape and its orientation in space. Starting from approach, excellent improvement of the spacecraft orbit relative to the comet is obtained from the optical measurements. Finally a (predicted) accuracy of less than 200 m (3σ) in spacecraft position relative to the centre of gravity of the comet can be reached at the pericentres during the close observation orbits.

Surface package delivery and relay. The surface package delivery dynamics and the related navigation aspects currently are studied /8/, in parallel to the development of the surface package design. The nominal scenario assumes that the surface package will be dynamically passive during its descent. It will be spun up by a separation spring eject device, and the spin axis will be aligned to the separation ΔV. The 'mother'-spacecraft orbits before and after separation are required to be safe.

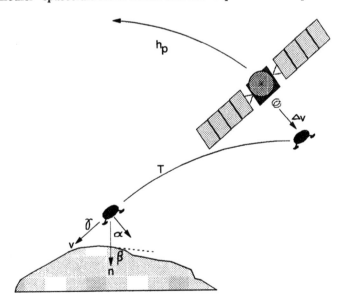

ΔV separation velocity

h_p pericentre of spacecraft orbit

T flight time of probe

V impact velocity

γ impact angle

α angle of attack

β surface slope

Fig. 4. Surface package delivery

The spacecraft position knowledge at pericentre has been estimated to be in the order of a few 100 m, therefore the pericentre altitude h_p has been chosen at 1 km. The separation conditions have been constrained by the requirement that the spin axis should be aligned as close as possible to the surface normal at impact (zero angle of attack α) allowing some unknown variation of the surface slope β, induced by the landing error. This was necessary to ease the probe damping and rebound design. In addition the spacecraft and probe orbits were both required to be elliptic. For the spacecraft this means that before probe separation several revolutions around the comet can be flown without manoeuvres, which is expected to lead to a very precise orbit knowledge from the landmark navigation, and thus to a good probe delivery accuracy. For the probe it means that in case of failures in the impact damping, the probe will not leave the cometary gravity field, and will eventually come to rest on the surface. The choice of the impact point was assumed not be restricted by the engineering, this means landing at the pole or the equator was to be possible. The separation spring energy (ΔV) can be chosen, but will have to be fixed before launch. Also the flight time of the probe from separation to impact ($T < 10$ hours) and the impact velocity ($V < 2$ m/s) will be constrained by the probe design. A vertical impact (velocity along surface normal, impact angle $\gamma = 90°$) would be desirable, but cannot be achieved in all cases, imposing all above conditions.

For a spring sizing for 0.3 m/s separation velocity, the spacecraft orbit before separation, and the time and direction of the surface package separation can be chosen such all above conditions are satisfied. A strategy has been constructed in which the package arrives at the surface with a spin axis orientation along the surface normal and with minimum vertical and horizontal (maximum impact angle) velocity components. Figure 5 shows the spacecraft and probe orbits for the extreme cases of comet size and density. The very small spring sizing is induced by the rebound escape constraint for the minimum gravity comet. If impact damping can be ascertained, higher separation velocities are possible, which then shorten the descent flight time T for the large comet cases, and thus improve the landing accuracy. The error analysis for the surface package delivery is ongoing /8/. After probe separation, the spacecraft is manoeuvred to an orbit which is best suited to receive the data transmitted from the surface package and to relay them to Earth. Originally only a few hours of surface package operations were foreseen. The scientific requirements now seem to evolve to a surface package operations time of at least 3 to 5 (Earth-)days.

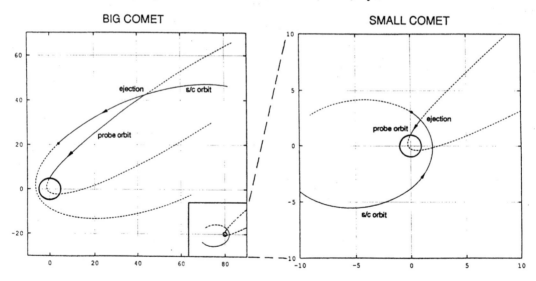

Fig. 5. Extreme surface package delivery orbits

Alternative scenarios with an additional spacecraft manoeuvre before the surface probe separation, which allows to align the impact velocity to the surface normal, and scenarios with special separation devices which decouple the separation direction from the spin axis direction have been studied as well (ref. /7/). The second surface package is assumed to be delivered a few months after the first, following similar strategies, and possibly using scientific information from the first package for site selection.

Extended monitoring (through perihelion). After the end of the activities related to the surface package science, the spacecraft will spend at least 8 months in orbit around the comet until after perihelion passage. The science goals of this phase are to monitor the nucleus (in particular active regions), dust and gas jets, and to analyse dust, gas and plasma in the inner coma from the onset to peak activity. Spacecraft orbits will be selected according to these scientific goals and spacecraft safety considerations. Mission planning will therefore depend on the result of previous observations, such as the activity pattern of the comet.

SPACECRAFT

The system study ROSETTA spacecraft design /2/ has been derived from a three-axis stabilised communications spacecraft. The main *configuration* features, the 400N biliquid engine, tanks size, solar array mounting and basic structure, of such a 'standard' bus were surprisingly well suited for the deep space mission. To cope with the large variations in sun and Earth distance and pointing directions which the spacecraft has to pass before arriving at the comet and in the orbit around the comet, the ROSETTA spacecraft uses solar arrays with a newly developed type of Low Intensity Low Temperature (LILT) cells with the usual one degree of freedom mounting in two wings.

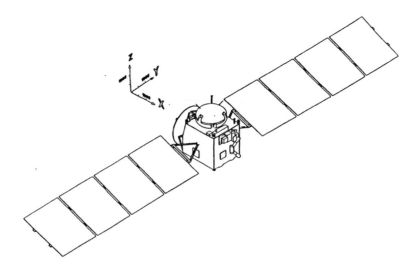

Fig. 6. System study spacecraft configuration

The 2 m high gain antenna dish is mounted on a two degrees of freedom mechanism. All payload and the navigation equipment (cameras, star sensors, gyros, accelerometers) are body fixed. This design allows to keep the sun pointing of the arrays and the Earth pointing of the antenna through all mission events which require pointing, namely main engine manoeuvres, observations of objects by the payload (asteroids, landmarks on comet surface), and surface package separation. The configuration is shown in figure 6. The thermal design uses external MLI insulation, radiators and electric heaters located where required. The extreme variations of input power from the solar arrays are handled by a special power point management. Attitude control will use 4N biliquid thrusters, reaction wheels, gyros and star and comet target tracking sensors. Communication will use an S-band command uplink and an X-band downlink. A medium gain antenna and two body mounted low gain antennae will support communications during the initial mission phase, during the Earth flybys and for emergency commanding. Communications with the surface package will be by means of omnidirectional VHF antennae. In the *Data Handling* system design particular attention has been given to the necessity of allowing autonomous spacecraft functioning also in cases without active ground intervention, for extended periods. The ground activity during these mission phases can be reduced to monitoring of the spacecraft health every two to three weeks. Special on board features for autonomous reconfiguration at the occurrence of a failure during the period without contact and reacquisition of ground control will be implemented. Also for mission periods under ground control the data handling system will provide an *On Board Master Schedule* function. This means a sequence of commands can be stored on board and sequentially modified and extended, for time tagged execution.

CURRENT STATUS AND EVOLUTION OF MISSION DESIGN

Actual spacecraft system development work, phase B of the project, will start in 1997. Preparations for phase B have been started, the selection of the payload, started with the issue of the Announcement of Opportunity, has to be completed before the phase B contract can be specified. Also the concepts of the surface packages, with respect to engineering and programmatic aspects, must be settled.

During the time before phase B, several key items of the mission will be studied in depth in the frame of the Technological Research Program (TRP) and also in further system level studies. With the evolution of the spacecraft design, the mission design will have to be adapted. E.g. the choice of the navigation camera systems, which is one of those item of detailed studies, will affect the near comet strategies. Also modifications in the spacecraft configuration, e.g. the mounting mechanisms of solar arrays and high gain antenna, may have an impact on the mission concepts. The near comet orbit planning and navigation will be further refined in the frame of a TRP study. In particular the surface package delivery concepts will be updated in accordance with the surface package design concepts which are expected to be finalised in this year.

The interplanetary orbit design, and consequently the catalogue of mission opportunities, may be primarily affected by an evolution of the ARIANE 5 performance by 2003, and the evolution of the spacecraft mass. A cycle of mission reoptimisations can be expected in the phase before the spacecraft design reaches a final stage of definition. E. g. it may be possible to suppress the manoeuvres planned during the planetary flyby phases at the cost of an increase in the total ΔV. Or it may also be desirable to advance the arrival date at the comet in some cases, using additional propellant allocation. This would allow to extend the period of near comet operations with modifications of the solar array and assuming availability of the NASA/DSN. Major work on the mission opportunities is planned for the next years. This in particular is intended to provide a complete list of asteroid flyby combinations for each of the three comet rendezvous missions. Based on this list ground observations of potential targets can be initiated and the targets can be selected on scientific reasons. It also cannot be excluded that other feasible missions, possibly to other comets, will still be found.

REFERENCES

1. ROSETTA Comet Rendezvous Mission, ESA SCI(93)7, Sept. 93

2. ROSETTA System Definition Study Rider 2, Matra Marconi Space, Contract Report CNSR/FF/128.93, August 1993

3. Horizon 2000, ESA SP-1070, Dec. 1984

4. G. Schwehm, M. Hechler, ROSETTA - 'ESA's Planetary Cometstone Mission, ESA Bulletin, Feb. 1994

5. M. Noton, Orbit Strategies and Navigation Near a Comet, ESA Journal, Vol 16/3, 1992, pp. 349-362

6. E. Gonzalez-Laguna, ROSETTA Navigation Near a Comet, MAS Working Paper No. 344, Sept. 1993, ESOC

7. V.V. Ivashkin, Comet Surface Science Package Delivery Dynamics, ESOC Contract 8165/89/D/IM, Final Report, GMVSA 2031/94, Madrid, June 1994

8. E. Gonzalez-Laguna, ROSETTA Surface Science Station Delivery Dynamics and Navigation, MAS Working Paper No. 351, ESOC, to appear

FINE STRUCTURES IN THE MIDDLE ATMOSPHERE AND THEIR ORIGIN

Proceedings of the C2.4 Meeting of COSPAR Scientific Commission C which was held during the Thirtieth COSPAR Scientific Assembly, Hamburg, Germany, 11–21 July 1994

Edited by

C. R. PHILBRICK

Department of Electrical Engineering, Pennysylvania State University, 121 Electrical Engineering East, University Park, PA 16802-2705

Adv. Space Res. Vol. 19, No. 1, pp. (1)139–(1)144, 1997
© 1997 COSPAR
Printed in Great Britain. All rights reserved
0273–1177/97 $17.00 + 0.00

Pergamon

PII: S0273-1177(96)00044-0

AERODYNAMICAL EFFECTS IN NUMBER DENSITY MEASUREMENTS IN THE LOWER THERMOSPHERE WITH THE CONE INSTRUMENT

F.-J. Lübken

Bonn University, Department of Physics, Nussallee 12, 53115 Bonn, Germany

ABSTRACT

We present measurements of total number densities in the lower thermosphere with a new instrument called 'CONE', which basically consists of an ionization gage. The electrodes of this new instrument are made of concentric grids with high transparency. A correction algorithm for aerodynamical effects ('ram correction') due to reflection of the incoming molecules on the CONE base structure is developed. Intercomparison of total number densities derived from CONE with data from falling spheres shows reasonable agreement in the overlapping altitude range.
©1997 COSPAR. All rights reserved

INTRODUCTION

It has been realized early in the history of rocket borne measurements in the upper atmosphere, that the density (or any other quantity) measured by an instrument on a rocket may be aerodynamically affected by the movement of the rocket /1,2,3/. The conversion of the raw data to ambient quantities taking into account this 'ram effect' is fairly simple, provided that the mean free path of the molecules is larger than the dimensions of the instrument ('free molecular flow conditions'). Because of this requirement the ram correction can only be applied at altitudes above approximately 85 to 90 km. We have earlier used a ram correction algorithm for our mass spectrometric measurements /4/ and, more recently, for our total number density measurements with the TOTAL instrument (a successor of CONE). In both instruments the molecules enter the sensitive region of the instrument (i. e. the ion source) through a small orifice and only after being accommodated to the temperature of the sensor ('closed geometry'). Intercomparisons of total number densities deduced from TOTAL with data from other techniques show reasonable agreement in the overlapping altitude range /5,6/. Except from total number densities, the TOTAL instrument measures small scale number density fluctuations in the altitude range 120 to 65 km in order to study turbulence /7/. For this purpose the time constant of the sensor is the most important instrumental parameter. Although the time constant of TOTAL is on the order of milliseconds only /5/, we decided to build a new ionization gauge with an even smaller time constant. This new instrument, called CONE (Combined Electron and Neutral Sensor), has an 'open' design, that is the molecules enter the ion source region without hitting any walls, at least in the ideal case. The sensor is 'open' because the electrodes of the ionization gauge consist of spherical grids with high transparency. A detailed description of the CONE sensor is presented in /8/.

In this paper we will concentrate on the absolute number density measurements of CONE above approximately 85 km. Despite the fact that the electrodes of CONE are nearly transparent, we cannot totally neglect aerodynamic effects. The reason for this is that the electrodes are mounted on a pyramidal structure which acts as a reflector for the incoming molecules. In the next section we will analyse the density enhancement in front of this pyramidal structure and we will develop a ram correction algorithm. CONE has successfully been flown twice during the SCALE campaign in 1993. In the last section we will show results from these flights and will compare our measurements with data from another technique (falling sphere).

AERODYNAMICAL EFFECTS UNDER FREE-MOLECULAR CONDITIONS

The grids and the filament of the CONE instrument are mounted on a structure which acts as a reflector for the incoming particles and causes an enhancement of the number density in front of the structure. We will now describe a correction algorithm for this 'ram effect'. The main assumptions made in this section are /1/:

- Free molecular flow conditions: the stream of molecules impinging on the instrument must not interact with the stream of molecules diffusively reflected from the instrument surface.

- The incoming molecules are accommodated to the temperature of the surface and are then reflected diffusely (i. e. equally likely in all directions), exhibiting a Maxwellian velocity distribution according to the temperature of the reflecting surface. In other words, the 'accommodation coefficient' is assumed to be equal one. This assumption is reasonable for our case (see page 168 ff in /1/) and is a posteriori justified when notifying the good agreement with the falling sphere densities (see later).

The number of particles impinging per unit time on a surface with unit area which is moving with a velocity v_R and an angle of attack α relative to the ambient gas (assumed to be at rest) is given by all particles having a large enough velocity component towards the surface. This number can be evaluated from the Maxwellian velocity distribution according to the temperature of the incoming gas T_i (i. e. the ambient atmosphere). The result is (see /1/ for more details):

$$\Phi_i^\circ = n_i \sqrt{\frac{kT_i}{2\pi \overline{m}}} \left\{ e^{-S^2} + \sqrt{\pi} \cdot S \cdot [1 + erf(S)] \right\} = \frac{1}{4} n_i \overline{c_i} \tag{1}$$

where

Φ_i° is the flux of incoming molecules to the surface,
n_i is the number density of the incoming molecules,
$S = |\vec{v}_R| \cdot \cos\alpha \, / \sqrt{2kT_i/\overline{m}}$
$erf(S) = (2/\sqrt{\pi}) \cdot \int_0^S e^{-x^2} \, dx$ is the error function,
$\overline{m}$ is the mean mass of the molecules,
α is the angle between the rocket velocity vector, $\vec{v}_R$, and the normal to the surface of CONE. This angle is obtained from the rocket trajectory and from the orientation of the payload during flight.
$\overline{c_i} = 2\sqrt{\frac{2kT_i}{\pi \overline{m}}}$ is the mean random speed of the incoming molecules.

The flux of reflected molecules Φ_r° is given by the same formula as above, but for $v_R = 0$:

$$\Phi_r^\circ = \frac{1}{4} n_r^\circ \overline{c_r} = n_r^\circ \sqrt{\frac{kT_r}{2\pi \overline{m}}} \tag{2}$$

where
$\overline{c_r} = 2\sqrt{\frac{2kT_r}{\pi \overline{m}}}$ is the mean random speed of the reflected molecules,
n_r° is the number density of the reflected molecules immediately above the surface (we have added an index 'o' in order to distinguish n_r° from $n_r(R, \vartheta)$ which is the number density at a point (R, ϑ) above the surface).

Since the flux of incoming particles must equal the flux of reflected particles, i. e. $\Phi_i^\circ = \Phi_r^\circ$, we get (by inserting the equations from above) a relationship between the number density in the incoming and reflected stream of molecules:

$$n_r^\circ = n_i \sqrt{\frac{T_i}{T_r}} \left\{ e^{-S^2} + \sqrt{\pi} \cdot S \cdot [1 + erf(S)] \right\} \tag{3}$$

This is the relationship needed to obtain the ambient number densities (n_i) from the number densities measured in a 'closed instrument', as for example the TOTAL instrument /5/.

In order to evaluate the density enhancement in front of the CONE structure due to the flux of reflected molecules, we need to specify the angular dependence of the outgoing flux. Since all directions above the surface are equally likely, the flux from a unit area $dx\,dy$ into a given direction is proportional to the projected area $dx\,dy\cos\vartheta$ seen from that direction (see Figure 1):

$$\phi_r(R,\vartheta) \propto \phi_r^0 \cos\vartheta \cdot f(R) \tag{4}$$

The proportionality factor and the function $f(R)$ are given by the requirement, that the number of particles flowing through a half sphere with radius R centered around the unit area $dx\,dy$ must equal the number of particles leaving that area (all per unit time):

$$\int_0^{\pi/2}\int_0^{2\pi} \phi_r(R,\vartheta)R^2\sin\vartheta d\vartheta d\phi = \phi_r^0 dx\,dy \tag{5}$$

Inserting equation 4 and evaluating the integral gives:

$$\Phi_r(R,\vartheta) = \Phi_r^0 \frac{1}{\pi R^2}\cos\vartheta dx\,dy \tag{6}$$

The number density enhancement due to reflection from this unit area $dx\,dy$ is given by the flux of reflected molecules divided by the mean speed: $dn_r(R,\vartheta) = \Phi_r(R,\vartheta)/\overline{c_r}$. Inserting equation 2 and 6 gives:

$$dn_r(R,\vartheta) = \frac{\Phi_r(R,\vartheta)}{\overline{c_r}} = \frac{1}{4}n_r^0\frac{1}{\pi R^2}\cos\vartheta\,dx\,dy \tag{7}$$

Expressing n_r^0 by n_i (equation 3) gives:

$$dn_r(R,\vartheta) = \frac{1}{4\pi R^2}\cos\vartheta\,dx\,dy\,n_i\sqrt{\frac{T_i}{T_r}}\left\{e^{-S^2} + \sqrt{\pi}\cdot S\cdot[1 + erf(S)]\right\} \tag{8}$$

Equation 8 expresses the density enhancement $dn_r(R,\vartheta)$ due to reflection of molecules from the area $dx\,dy$ at a point with distance R and elevation angle ϑ above that area. We can now apply equation 8 to our CONE structure. We need to consider, however, that the reflecting surface in CONE is of pyramidal shape (see Figure 2). We therefore first evaluate the contribution of a concentric ring $dx\,dy = 2\pi\sin\eta s ds$ to the enhancement in a distance h from the top of the pyramidal structure, and than integrate over the whole pyramidal surface, i. e. from $s = 0$ to $s = \ell$ (ℓ = length of the pyramid edge). From geometrical considerations (cosine and sine law) we get the expressions needed for R^2 and $\cos\vartheta$ in equation 8:

$$R^2 = h^2 + s^2 + 2hs\cos\eta \tag{9}$$

$$\cos\vartheta = \frac{h}{R}\sin\eta = \frac{h\sin\eta}{\sqrt{h^2 + s^2 + 2hs\cos\eta}} \tag{10}$$

Inserting and integrating over all dn_r gives:

$$n_r = \int dn_r = \frac{1}{4\pi}n_r^0\int_0^\ell\left(\frac{1}{h^2 + s^2 + 2hs\cos\eta}\right)\cdot\left(\frac{h}{R}\sin\eta\right)\cdot 2\pi\sin\eta s\,ds \tag{11}$$

$$= \frac{1}{2}n_r^0 h\sin^2\eta\int_0^\ell\left(h^2 + s^2 + 2hs\cos\eta\right)^{-3/2} s\,ds \tag{12}$$

$$= \frac{1}{2}n_r^0\left\{1 - \frac{1 + \epsilon\cos\eta}{\sqrt{\epsilon^2 + 2\epsilon\cos\eta + 1}}\right\} \tag{13}$$

where we have introduced the abbreviation $\epsilon = \ell/h$. We can easily see that equation 11 makes sense for special cases: immediately above a flat surface ($\eta = 90°$; $\epsilon \to \infty$), we get $n_r = n_r^o$. Very far away from the surface ($\epsilon \to 0$) we get $n_r = 0$, as it should be. In the CONE sensor the geometrical conditions are such that $\eta = 80°$ and $\epsilon = 0.75$. This results in a geometrical factor (equation 11) of 0.081.

The so called 'ram factor', i. e. the factor by which the density in front of the CONE structure is enhanced, is defined as $(n_i + n_r)/n_i$. Inserting equations 3 and 11 gives:

$$\frac{n_i + n_r}{n_i} = 1 + \frac{n_r}{n_i} = 1 + \sqrt{\frac{T_i}{T_r}} \left\{ e^{-S^2} + \sqrt{\pi} \cdot S \cdot [1 + erf(S)] \right\} \cdot \left\{ \frac{1}{\sin \eta} \left(1 - \frac{1 + \epsilon \cos \eta}{\sqrt{\epsilon^2 + 2\epsilon \cos \eta + 1}} \right) \right\} \tag{14}$$

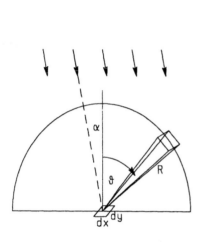

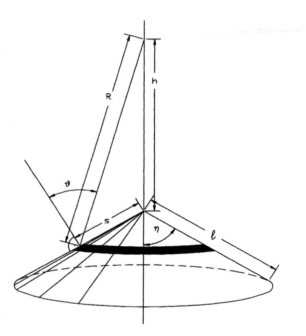

Figure 1: Flux through a solid angle due to reflection of molecules from a unit area $dx\ dy$ (see text).

Figure 2: Geometrical conditions at the base of the CONE structure (the variables are explained in the text).

We can know calculate the number density enhancement in front of the CONE structure and thus correct the measured densities (i. e. $n_i + n_r$) to the ambient number densities (i. e. n_i), provided the quantities on the r. h. s. of equation 14 are given. This is true for all quantities, except for the ambient temperature T_i. Since the temperature profile $T_i(z)$ is obtained by integrating $n_i(z)$ and, on the other hand, $T_i(z)$ is needed in equation 14 in order to obtain $n_i(z)$, we choose an iterative procedure: as a start we use the temperature profile from CIRA-1986 /9/ and derive $n_i(z)$. This density profile is used to deduce the temperature profile which is then input in equation 14 to obtain a more realistic density profile, and so on (the unknown 'start temperature' at the top of the profile is again taken from CIRA-1986 ; see e. g. /5/ for more details). The procedure converges after a few iterations.

APPLICATION TO THE CONE INSTRUMENT

We will now apply the formalism developed in the preceeding section to the rocket flight labeled 'SCT03' launched at 22:23:00 UT on 28.7.1993 from the Andøya Rocket Range (69°N,16°E) in the scope of the SCALE campaign in Northern Norway (SCAttering Layer Experiment). The CONE instrument was mounted on the TURBO sounding rocket together with other experiments. Shortly before (16 minutes) and after (15 minutes) flight SCT03 a so called 'falling sphere' was launched which gives total densities, temperatures, and horizontal winds in the altitude range 95 to 35 km. We should note that this technique gives profiles which are smoothed over a few kilometers in the upper mesosphere /6/.

In Figure 3 the ram factor defined in equation 14 is shown for the conditions in flight SCT03. For comparison we also show the ram factor for the TOTAL instrument, as defined in equation 3. As can be seen from this Figure the ram effect is much smaller in the CONE instrument compared to TOTAL. Still, the number density enhancement can be as large as $\approx$ 60 % around 95 km. Since the mean free path as derived from the falling sphere densities is 73 mm, 26 mm, and 8 mm at an altitude of 94 km, 90 km, and 85 km, respectively, and since the distance between the pyramidal structure and the ionizing region is approximately 10 mm, we can apply the ram correction to altitudes above $\sim$ 85 km. The upper altitude limit of the density measurements is given by the sensitivity of CONE and by an offset on the electrometer zeropoint caused by photoelectrons. We choose to start our measurements in an altitude where the uncertainty due to this offset has reduced to below 5% of the total signal, that is below 108 km.

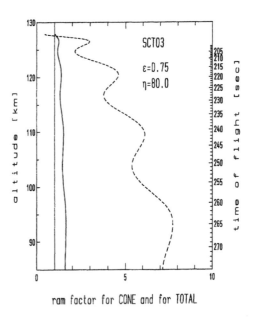

ram factor for CONE and for TOTAL

Fig. 3: Ram factor for CONE (solid line) and TOTAL (dashed line) as defined in equation 14 and 3, respectively.

In Figure 4 the ratio of the number densities measured by CONE during flight SCT03 and by the two falling sphere flights are shown relative to CIRA-1986. As can be seen from this Figure, there is reasonable agreement between the falling sphere and the CONE densities. The remaining difference is presumably due to the more complicated geometrical shape of the ionization region, which may at least partly be located closer to the surface as assumed in our derivation.

In Figure 5 the temperature profiles derived from the density profiles in Figure 4 are shown. As mentioned earlier, the 'start temperature' for the CONE profile is taken from CIRA-1986, whereas the start temperatures for the two falling sphere profiles are taken from a smoothed CONE profile. As can be seen from Figure 5 there is good agreement between the CONE and the falling sphere temperature profiles. Due to its better hight resolution CONE shows more structure than the falling spheres. It is interesting to note that the CONE profile shows a temperature gradient close to an adiabatic lapse rate between approximately 98 and 100 km. Compared to CIRA-1986 the mesopause is significantly colder in all three profiles.

REFERENCES

1. G. N. Patterson, *Molecular Flow of Gases*, John Wiley & Sons, New York, 1956.

2. R. Horowitz and H. E. LaGow, Upper air pressure and density measurements from 90 to 220 km with the Viking 7 rocket, *J. Geophys. Res.* 62, 57-78, 1957.

3. E. J. Schaefer and M. H. Nichols, Upper air neutral composition measurements by a mass

spectrometer, *J. Geophys. Res.* 69, 4649 – 4660, 1964.

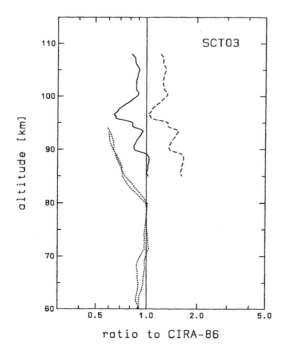 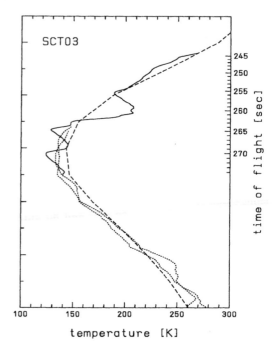

Fig. 4: Ratio of densities measured by CONE (solid line: ram corrected, dashed line: raw data) and falling spheres (dotted lines) relative to CIRA-1986.

Fig. 5: Temperatures deduced from the density profiles in Figure 4. For comparison the CIRA-1986 profile for July and 70°N is also shown (dashed line).

4. U. von Zahn, F.-J. Lübken and Ch. Pütz, 'BUGATTI' experiments: mass spectrometric studies of lower thermosphere eddy mixing and turbulence, *J. Geophys. Res.* 95, 7443–7465, 1990.

5. W. Hillert, F.-J. Lübken and G. Lehmacher, Neutral air turbulence during DYANA as measured by the TOTAL instrument, *J. Atmos. Terr. Phys.* , in press, 1994.

6. F.-J. Lübken, W. Hillert, G. Lehmacher, U. von Zahn, M. Bittner, D. Offermann, F. Schmidlin, A. Hauchecorne, M. Mourier and P. Czechowsky, Intercomparison of density and temperature profiles obtained by lidar, ionisation gauges, falling spheres, datasondes and radiosondes during the DYANA campaign, *J. Atmos. Terr. Phys.* , in press, 1994.

7. F.J. Lübken, W. Hillert, G. Lehmacher and U. von Zahn, Experiments revealing small impact of turbulence on the energy budget of the mesosphere and lower thermosphere, *J. Geophys. Res.* 98, 20,369–20,384, 1993.

8. J. Giebeler, F.-J. Lübken and M. Nägele, CONE - a new sensor for in-situ observations of neutral and plasma density fluctuations, *Proceedings, ESA-SP-355*, Montreux, Switzerland, 311 – 318, 1993.

9. E. L. Fleming, S. Chandra, J. J. Barnett and M. Corney, Zonal mean temperature, pressure, zonal wind and geopotential height as functions of latitude, *Adv. Space Res.*, 10,#12, 11–59, 1990.

Acknowledgements: *I thank Jochen Giebeler for performing the CONE measurements and Rolf Becker for processing the falling sphere data. I appreciate the efforts of Hinnerk Baumann who has built the CONE sensors. I thank the DLR-MORABA, the NDRE, and the Andøya Rocket Range staff for their excellent work. This study was supported by the BMFT, Bonn, under grant 01OE88027.*

Adv. Space Res. Vol. 19, No. 1, pp. (1)145–(1)148, 1997
© 1997 COSPAR
Printed in Great Britain. All rights reserved
0273–1177/97 $17.00 + 0.00

 Pergamon

PII: S0273-1177(96)00045-2

GRAVITY WAVE SIGNATURE SIMULTANEOUSLY OBSERVED IN THE OXYGEN ATOM AND ELECTRON DENSITY PROFILES IN THE LOWER THERMOSPHERE

T. Imamura, K. Kita, N. Iwagami and T. Ogawa

Department of Earth and Planetary Physics, Graduate School of Science, University of Tokyo, Bunkyo-ku, Tokyo 113, Japan

ABSTRACT

The fine structures of lower thermospheric atomic oxygen and electron density profiles obtained by a rocket experiment were investigated, assuming that these structures are due to a quasi-monochromatic internal gravity wave. The height variation of the horizontal wind amplitude is similar to that of the intrinsic horizontal phase speed below ∼105 km, implying an evidence of saturation effects. At higher altitudes, the dissipation due to molecular diffusion was suggested. The wavy variation of the intrinsic horizontal phase speed with height can be explained by the mean wind modulation due to an atmospheric tide. The horizontal phase velocity vector inferred is interpreted as a consequence of selective transmission of internal gravity waves in lower atmosphere.
©1997 COSPAR. All rights reserved

OBSERVATION

The observation was carried out by the sounding rocket S310.21 flown from Uchinoura (31°15'N, 131°05'E) at 1200UT on 28 January 1992. The atomic oxygen density was measured with an improved resonance fluorescence technique /1/. The electron density was retrieved from the height distribution of the magnetic field intensity of VLF waves from broadcasting stations and the DC probe collecting electrons /2/.

Figures 1 and 2 show the measured atomic oxygen and electron densities, respectively. They were measured simultaneously in the rocket ascent. Both of the profiles have several fine structures, which can be attributed to atmospheric waves. Atomic oxygen can be used as a tracer of atmospheric motions because its photochemical time constant of several days /3/ is much longer than the typical periods of gravity waves in this region /4/. Though the removal time constant of electron is short in the case that molecular ions are dominant, electron density also responds to atmospheric motions due to the response of metallic ions whose removal time constants are long enough (>100 days). In the present paper these fine structures are assumed to be created by a quasi-monochromatic internal gravity wave.

CHARACTERISTICS OF THE WAVE

The horizontal broken lines in Figure 1 indicate the positions of the peaks and the dips of the wavelike structures. The interval between the adjoining peak and dip represents a half of the vertical wavelength λ_z at each altitude except at ∼92 km/105 km where the position of the dip/peak cannot be determined. The distinct structure around 104 km is considered as a spurious peak created by the large dip just below this structure, because the density response equation /5,6,7/ predicts that the density response disappears just above this altitude where the phase reversal occurs. The height profile of the intrinsic horizontal phase speed $|C - \hat{\mathbf{k}}_h \cdot \bar{\mathbf{u}}|$ is obtained from λ_z using the dispersion relation, where C is the horizontal phase speed, $\hat{\mathbf{k}}_h$ is the unit vector of horizontal phase velocity and $\bar{\mathbf{u}}$ is the horizontal mean wind.

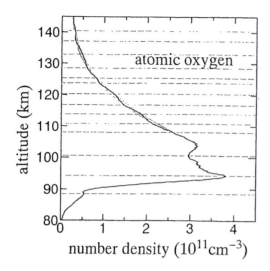

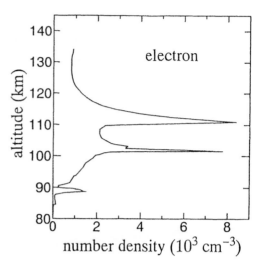

Fig. 1. Measured atomic oxygen profile. Horizontal broken lines indicate the peak and dip heights. Envelopes are drawn to distinguish wavy structures.

Fig. 2. Measured electrn density profile showing distinct layers.

An initial profile of atomic oxygen without gravity wave perturbations was inferred as in Figure 3, by smoothing the observed profile above ~105 km and by a number of numerical simulations for many model profiles using the density response equation below ~105 km. This initial profile responds to a gravity wave almost in the same way as the observed structures according to the numerical calculation with the vertical wavelengths inferred above (see Figure 4). The amplitude for horizontal wind, u'. can be obtained from the amplitude for atomic oxygen density fluctuation using the density response equation.

Fig. 5. illustrates the height profiles of u' and $|C - \hat{\mathbf{k}}_h \cdot \bar{\mathbf{u}}|$ obtained. Since u' is close to $|C - \hat{\mathbf{k}}_h \cdot \bar{\mathbf{u}}|$ considering error bars below ~105 km, the observed gravity wave is likely to be saturated and breaking due to convective instability rather than molecular viscosity at these altitudes. The expected distortion of density response due to the localized intense turbulence associated with saturation effect is discussed elsewhere /8/. Above ~105 km the wave dissipates mainly due to molecular viscosity because an energy loss is implied from $u' < |C - \hat{\mathbf{k}}_h \cdot \bar{\mathbf{u}}|$ and the growth rate of u' with height. It should be noted that this critical altitude of ~105 km has an ambiguity of vertical wavelength (~10 km). If the wave packet distributes over the height range from 80 km to 150 km and C and $\hat{\mathbf{k}}_h$ are almost constant, the structured profile of $|C - \hat{\mathbf{k}}_h \cdot \bar{\mathbf{u}}|$ would suggest that the mean wind $\bar{\mathbf{u}}$ is modulated by an atmospheric tide. The wavelike variation of u' with height below ~105 km would be a result of wave-tidal interaction suggested by Fritts and Vincent /9/, who have argued that the amplitude of a gravity wave is limited by the local mean wind modulated by tidal motions. The wavy structure of u' above ~105 km would be a result of the wavelike variation of $|C - \hat{\mathbf{k}}_h \cdot \bar{\mathbf{u}}|$, considering the WKB solution of the Taylor-Goldstein equation describing the amplitude and vertical structure of gravity waves.

The propagation direction of the wave can be determined from both atomic oxygen and electron profiles based on the fact that the response of ionic species to gravity waves is different from that of neutral species depending on the propagation direction. For example, for waves propageting eastward, the region in which ionic layers are formed is located at the phase of about $-\pi/2 \sim \pi/2$ downward from the peak positions of background density fluctuation. For westward propagation, it is located at $\pi/2 \sim 3\pi/2$. Using a simple model, the horizontal phase velocity vector was inferred

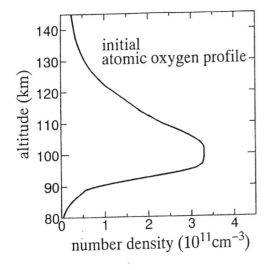

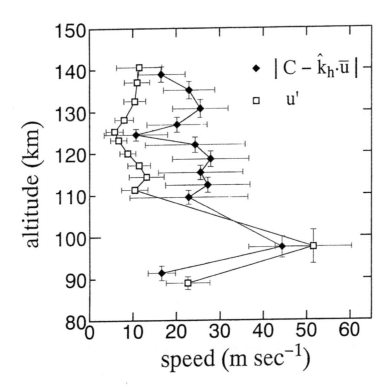

Fig. 3. Initial atomic oxygen density profile inferred.

Fig. 4. Observed atomic oxygen profile (solid) and simulated profile at around 100 km for gravity wave perturbation with $\lambda_z = 12.8$ km (dotted).

Fig. 5. Height profiles of intrinsic horizontal phase speed (solid diamonds) and horizontal wind amplitude (open squares).

to be 20~45 m s^{-1} and nearly westward /8/.

The linear saturation theory /10/ shows that gravity waves cannot pass through the critical level where $|C - \hat{k}_h \cdot \bar{u}| = 0$. According to the radiozonde and meteorological rocket observations by Japan Meteorological Agency, the critical level did not exist below ~58 km altitude for waves with the horizontal phase velocity inferred above. The CIRA1986 model atmosphere /11/ suggests that the mean zonal wind is eastward below 100 km altitude at 31°N in January or February, implying that the wave packet can also pass through the mesosphere. These can be interpreted by the selective transmission of gravity waves in lower atmosphere. The meteor radar observations /12/ suggests that the horizontal phase velocity vector inferred in this study is typical in the lower thermosphere at around the observation site in winter.

Acknowledgement The authors wish to thank Prof. M. Mambo, Drs. T. Okada and T. Fukami for providing the electron density data. Thanks are also due to Dr. T. Nakamura, Dr. M. D. Yamanaka, Prof. S. Fukao and Dr. M. Yamamoto for helpful discussions. The S310.21 rocket experiment was carried out under the research program of the Institute of Space and Astronautical Science, Japanese Ministry of Education, Science and Culture.

REFERENCES

1. K. Kita, T. Imamura, N. Iwagami and T. Ogawa, Rocket observation of oxygen atom and night airglow 1. measurement of oxygen atom concentration with an improved resonance fluorescence technique, submitted to *Ann. Geophysicae*.

2. M. Mambo, T. Okada and T. Fukami, private communication (1994).

3. R. R. Garcia, and S. Solomon, The effect of breaking gravity waves on the dynamics and chemical composition of the mesosphere and lower thermosphere, *J. Geophys. Res.* <u>90</u>, 3850-3868 (1985).

4. A. H. Manson, Gravity wave horizontal and vertical wavelengths: an update of measurements in the mesopause region (~80-100 km), *J. Atmos. Sci.* <u>47</u>, 2765-2773 (1990).

5. Y. T. Chiu, and B. K. Ching, The response of atmospheric and lower ionospheric layer structures to gravity waves, *Geophys. Res. Lett.* <u>5</u>, 539-542 (1978).

6. C. S. Gardner, and J. D. Shelton, Density response of neutral atmospheric layers to gravity wave perturbations, *J. Geophys. Res.* <u>90</u>, 1745-1754 (1985).

7. J. Weinstock, Theory of the interaction of gravity waves with $O_2(^1\Sigma)$ airglow, *J. Geophys. Res.* <u>83</u>, 5175-5185 (1978).

8. T. Imamura, K. Kita, N. Iwagami and T. Ogawa, Gravity wave breaking at 90-140 km derived from atomic oxygen and electron density profiles, submitted to *J. Geophys. Res.*

9. D. C. Fritts, and R. A. Vincent, Mesospheric momentum flux studies at Adelaide, Australia: observations and a gravity wave-tidal interaction model, *J. Atmos. Sci.* <u>44</u>, 605-619 (1987).

10. R. S. Lindzen, Turbulence and stress owing to gravity wave and tidal breakdown, *J. Geophys. Res.* <u>86</u>, 9707-9714 (1981).

11. E. L. Fleming, Zonal mean temperature, pressure, zonal wind and geopotential height as functions of latitude, *Adv. Space Res.* <u>10</u> (12), 11-59 (1990).

12. M. Yamamoto, T. Tsuda and S. Kato, Gravity waves observed by the Kyoto meteor radar in 1983-1985, *J. Atmos. Terr. Phys.* <u>48</u>, 597-603 (1986).

Pergamon

PII: S0273-1177(96)00046-4

Adv. Space Res. Vol. 19, No. 1, pp. (1)149–(1)158, 1997
© 1997 COSPAR
Printed in Great Britain. All rights reserved
0273–1177/97 $17.00 + 0.00

THIN ION LAYERS IN THE HIGH-LATITUDE LOWER IONOSPHERE

S. Kirkwood

Swedish Institute of Space Physics, Box 812, S-981-28 Kiruna, Sweden

ABSTRACT

Thin ion layers in the high-latitude lower ionosphere can be produced by the action of wind-shears and electric fields on metallic and other ions and, perhaps, by other processes. Modelling studies have been used to study the two former mechanisms and it has been found that electric field effects will usually dominate those of winds at auroral latitudes, producing layers at altitudes as low as 90 km. When electric fields are very small, however, layers should be produced by tidal winds or gravity waves, as they are at other latitudes. A considerable number of observations of thin ion layers have been made during the 10 years of operation of the EISCAT incoherent scatter radar. Most, but not all, of these layers are likely to be composed of metallic ions. Many fit well with model calculations of formation by electric field action, others are better explained as a result of wind shear. The average seasonal and daily variations of the occurrence frequency and altitude distribution of the layers can be well explained by a combination of the effects of winds and electric fields. However, a number of puzzling features still remain unexplained, for example the correlation of metallic ion layers with layers of neutral atoms and the extreme thinness of some of the layers. The successes and failures of the theories of their formation will be discussed.
©1997 COSPAR. All rights reserved

INTRODUCTION

The occurrence of thin ion layers in the high-latitude E-region has long been inferred from ionosonde observations. Comrehensive studies of such 'sporadic-E' layers using the ionosonde in Sodankyla, Northern Finland can for example be found in /1,2,3,4,5/. Classic sequential Es, slowly descending layers of high electron density can often be seen , particularly on summer mornings and afternoons. The same signatures seen at lower latitudes are known from sounding rocket investigations to be caused by thin layers of metallic ions, thought to be formed by the action of tidal wind-shears. (For a review of mid-latitude Es, see /6/) Since the same mechanisms could operate at high-latitudes (albeit less effectively) it was not too surprising that such layers were also seen there. Sporadic-E during night hours is often due to ionisation of the major atmospheric constituents by auroral particle prcipitation but there were indications from ionosonde studies that at least some night-time sporadic-E was also due to thin layers of long-lived ions /5/. However , it was only with the advent of incoherent-scatter radar mesurements at high latitudes, with height resolution of a few km or less, that the full diversity of thin ion layers occurring there could be discovered. Studies of ion-composition in high-latitude Es layers have been made / 7,8 / showing, indeed, that metallic ions are responsible for many thin layers. IS radar also offered the possibility to test the 'accepted' wind-shear formation mechanisms for thin ion layers and to nighlight it's shortcomings. Many of the features of high-latitude thin ion layers have turned out to be explicable by a combination of electric field and wind-shear effects but there remain a number of unexplained aspects - particularly the extreme thinness and symmetry (in altitude) of the layers even at the lowest altitudes and their common correlation with neutral metallic layers /9,10,11/.

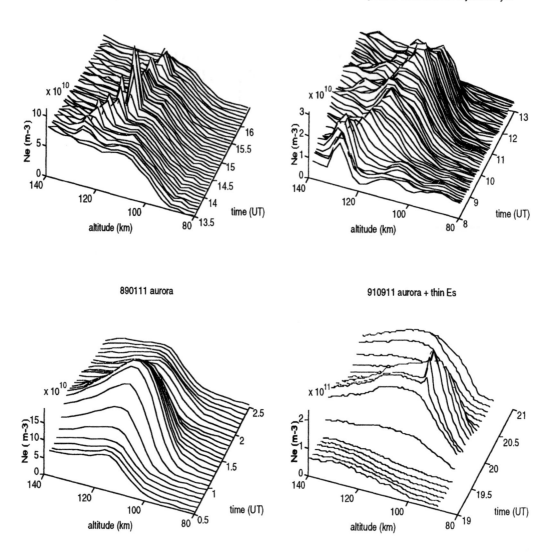

Figure 1. Typical thin ion layers observed by the EISCAT IS radar, at the dates and times indicated. The first three examples are measured with 3 km height resolution, the last example (lower RHS) with 600 m resolution.

The aim of this report is to demonstrate the characteristics of high-latitude thin-ion layers, to describe the electric-field and wind-shear theories which have been proposed to explain their formation, and to examine how successfull those theories are in explaining the observations.

TYPICAL THIN ION LAYERS AT 70°N, 19°E.

Figure 1 shows a number of examples of typical thin ion layers, measured with the EISCAT UHF incoherent scatter radar situated near Tromsø in northern Norway. Some of the examples are from the radar 'Common Program' database, and have a height resolution of 2-3 km. Other examples are from special measurement campaigns and have much better height resolution, 600 m. They have been selected from a database including all of the 'Common Program 1' observations between 1989 and 1992 (31 days) , plus a total of 20 evenings of special measurements during August 1990 and September 1991.

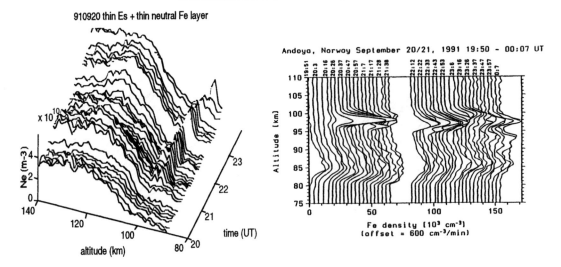

Figure 2. Typical low-altitude night-time Es layer observed by EISCAT (LHS, 600 m height resolttion) and correlated thin layer of neutral Fe atoms measured by lidar over Andøya, 130 km west of the radar site.

The first example in Figure 1 and the first 2 h of the second example, afternoon thin Es and morning thin Es, show the type of Es which could easily be identified by ionosonde - a slowly descending thin layer, with thickness substantially less than the atmospheric scale height. The second exapmle also shows, between 11-13 UT, a sightly broader layer of ionisation, apparently 'fed' by the Es layer and a number of weaker descending 'clouds' of ions, which is typical of winter 'daytime' /12/. (At high latitudes winter 'daytime' means that the sun is at best a few degrees above the horizon). The composition of these broad winter layers is as yet unknown. The next example illustrates the common situation during auroral precipitation - an electron density profile with a sharp lower border but a topside gradient comparable with or longer than the atmospheric scale height. This profile is caused by precipitation of energetic particles and will not, in fact , be considered a thin ion layer for the purposes of this report. The following example illustrates an aurorally-produced E-layer together with a superimposed much thinner layer, most likely a metallic-ion layer. This combination is not uncommon but would have been very difficult to detect with an ionosonde. Figure 2 shows another typical thin ion layer - a metallic ion layer at low altitude on a very quiet evening, without any auroral particle precipitation. This particular layer was sampled by a rocket-borne mass spectrometer and was found to be mainly Mg+ /8/. It was also closely associated with a neutral Fe layer, observed at almost the same geographical location, almost the same height and overlapping in time. The neutral layer is also shown in Figure 2.

Figure 3 shows some less typical thin ion layers. The last example in the figure is an unusually intense sporadic-E layer. Only 3 of the 80 layers in the data interval searched reached this intensity. The example in Figure 3 can also be seen to develop into a triple layer. Double layers are not uncommon, and sounding-rockets with better height resolution and better sensitivity to weak layers almost always see multiple layers (e.g./8/). However, the intensity , the rather large separation between the two most intense layers and the appearence of a strong-enough third layer to be seen in IS measurements in this case are unusual. The third example in Figure 3 shows an upward migrating layer, the only example in the database where upward movement was seen for more than a few minutes. The second example shows a pulsating layer, varying cyclically in intensity as it descends, a unique example. The first example is a rather broad layer, several km thick but still with topside gradient less than the atmospheric scale height. This type of layer was seen on 4 occasions, during both summer and winter The origin of this type of layer is unclear, but they may have a similar explanation to the winter daytime layers (Figure 1, 2nd example). Certainly their morphology suggests that they may be formed by trasport of long lived ions, in a similar way to metallic-ion sporadic E layers.

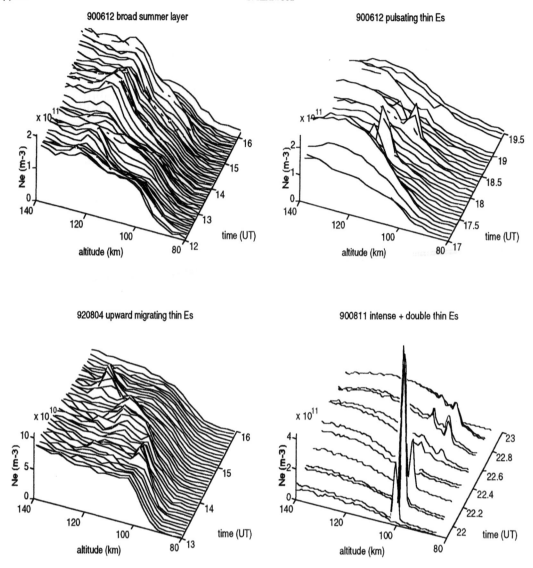

Figure 3. Unusual thin ion layers observed with EISCAT. The first 3 examples have 3 km height resolution , the fourth (lower RHS), 600 m.

STATISTICAL CHARACTERISTICS OF THIN ION LAYERS AT 70°N, 19°E

Figure 4 shows the distribution of the occurrence of Es layers in the IS observations as a function of altitude, time of day, season and the direction of the electric field. Only thin ion layers of the type identified as 'thin Es' in the earlier figures are included in these statistics, i.e. not those caused by auroral particle precipitation, and not those thicker than 5 km. Also, only the radar 'Common Program' observations have been included in order to have as uniform a coverage of time of day as possible. The number of observation hours included in this sample is rather small (753 h) compared to previous studies based on ionosonde observations. However, it is sufficient to show the same statistical pattern for the 'classic' daytime Es in addition to shedding considerable light on the characterictics of night-time Es. Considering the time-of-day variation first, we see that Es are much more common between 18-02 UT (ca. 19-03 LT) than at other times of day. This could not be appreciated using ionosonde data because of the difficulties in separating auroral particle precipitation from true, thin Es. However, the Es layers at night are at lower altitudes than those seen during the day - below 110 km in the evening and mostly below 100 km after midnight. The 'classic'

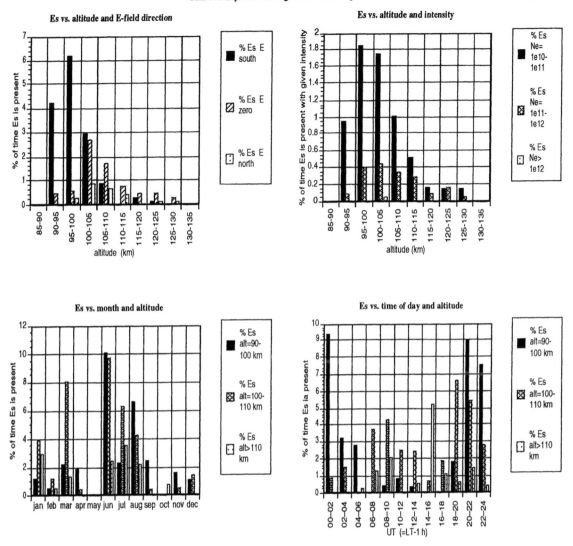

Figure 4. Statistical properties of thin-ion layers measured by EISCAT (based on 31 days of 'Common Program' measurements, 1989-1992). Only layers of the types identified as 'thin' in Figures 1,2 and 3 are included in the statistics, i.e only layers less than 5 km thick.

Es are seen in the daytime statistics mostly between 100-110 km in the pre-noon sector but at higher altitutes post-noon. This is in good agreement with the ionosonde statistics /1,2,3/.

If we next consider the intensity of the Es layers, we see that most layers are rather weak, with peak density less than 10^{11} m^{-3}. In practice, the peak densities may be a little higher than indicated here since the observed values are averages over a ca. 3 km height interval. There is perhaps a slight tendency for weaker layers at lower altitudes but this is not particularly marked. What is significant, is the complete absence of layers below 90 km and above 130 km. The third panel in Figure 4 shows the seasonal distribution of Es occurrence. A clear concentration in the summer months can be seen, in good agreement with ionosonde studies. It is noteworthy that this seasonal dependence is seen also in the low-altitude layers which are predominantly from the night-hours (top panel) and were not included in earlier statistics based on ionosonde studies. The same seasonal variation is seen in mid-latitude layers and no good explanation has yet been found /6/. There is no good correlation with the influx of meteors and, although seasonal variations in the tidal wind shears could be suggested as a reason, few measurements of the wind shears exist.

S. Kirkwood

simulation - Es formed by semidiurnal tide

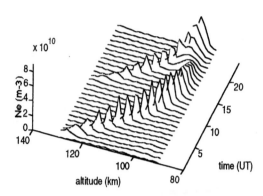

Figure 5. Simulation of Es layer formation by a semidiurnal tidal zonal wind (wind amplitude 60 m/s, vertical wavelength 40 km)

The fourth panel in Figure 4 shows the height distribution of Es for different electric field directions. Here it can be seen that by far the majority of the low-altitude (nightime) Es layers occurr for southward directed electric fields. (The E-W component of the electric field is generally much smaller than the N-S component, see below, so only the former has been used to sort the observations). Higher altitude Es layers are usually related to zero (< 3 mV/m) electric fields. This kind of comparison was impossible using ionosonde data and points to an earlier unrecognised importance for the electric field in Es formation.

THEORIES OF LAYER FORMATION

Two mechanisms which can gather long-lived metallic ions into thin layers are known to operate - wind shear as proposed by Whitehead /13/ and electric field, first described in detail by Nygrén et al /14/. Figure 5 shows model calculations of Es formation by a shear in the E-W wind associated with a semidiurnal tide. This tidal mode is known to be dominant in the lower E-region at high latitudes

simulation - Es formed by electric field which rotates from NW to S

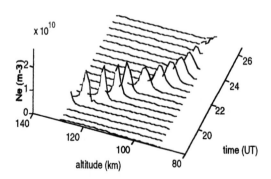

Figure 6. Simulation of Es layer formed by the action of an electric field which is initially directed NW , rotating towards S over the interval of the simulation. The electric field used is the average field shown in Figure 8, from 18-02 UT.

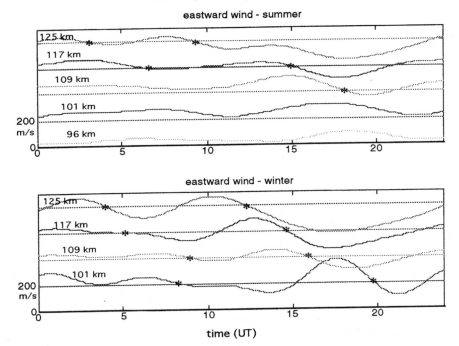

Figure 7. Daily variation of the zonal neutral wind over EISCAT . The figure shows the sum of the mean, 24 h, 12h, 8h and 6h components of the average wind variation from /15/. The * mark the positions of the convergent nodes , where thin Es layers could be formed by wind-shear.

with roughly the phase, amplitude and vertical wavelength used in the simulation. Clearly tidal wind shear could account for the morning and evening thin Es. However, electric fields provide a strong modification of this pattern, and as soon as the field strength exceeds a few mV/m completely dominate over the tidal effect /11/. An electric field directed between north and east stronger than a few mV/m would remove all convergence. A field directed between north and west would produce its own convergent null at some heights above about 105 km. In fact, an field rotating from northward to westward would produce a signature very similar to the tidal wind-shear (Fig. 6). An electric field with a southward component rapidly pushes Es layers below 105 km, and can even form an Es layer due to the rapidly decreasing ion drift speed at lower altitudes /10,11/. In order to be able to assess the cause of high-latitude Es layers we must know more about the wind-shears and the electric fields.

TIDES AND ELECTRIC FIELDS AT 70°N, 20°E

Figure 7 shows the average winds at a number of E-region heights, during summer and winter, from a comprehensive study of winds using the EISCAT IS radar /15/. The winds shown include the prevailing, 24 h, 12 h, 8h and 6h components from that study. The * mark positions of convergent nodes where Es layers should be formed by the wind shear mechanism. Clearly the combination of several components makes for a much less clear pattern of descending nodes than in the simulation in Figure 5. The afternoon node is clearest, and in good agreement with the typical afternoon thin Es in Figure 1. There is almost a descending node between 05-10 UT, which could explain the morning thin Es (as in Fig. 1, second example). However, there are three features which need further explanation. The first is that the afternoon wind shear is best developed in the winter, whereas the corresponding Es are much more common in the summer. The second is that , according to the statistics in Figure 4, the morning thin Es are just as common as the afternoon Es, despite the fact that the wind-shear node is much better developed in the afternoon. The third is that there is clearly no wind-shear node which can account for the night-time Es.

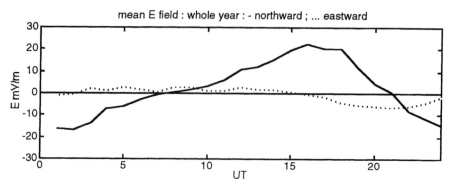

Figure 8. Mean electric field as a function of time of day for the 31 days of 'Common Program 1' operations with the EISCAT radar 1989-1992.

Figure 8 shows the average electric field (as measured by the IS radar) for all of the observation days used for the Es statistics. The variation of electric field direction over the day is not significantly different between summer and winter, although the electric field strength is on average, 50% higher during the winter days than the summer. The field strength does vary considerably from day to day, however, and figure 9 shows how often the field is negligible (relative to the tidal wind amplitudes). From this information we can see that the night-time Es layers , and their slightly higher altitudes in the early evening, can be well explained by the typically north-westward, rotating to southward, electric fields between 18-05 UT. The morning Es can be formed by tidal wind shear since the electric field is very small 40-60% of the time between 04-12 UT. On the other hand, formation of Es by the better developed afternoon wind shear will often be hindered by a strong electric field (the field is small only 20 % of the time between 15-02 UT

SUCCESSES AND FAILURES OF THE THEORIES

The combination of tidal wind shear and electric field seems to be enough to explain the main features of the altitude distribution and daily variation. The absence of Es above 130 km may be explained by the increasing dominance of the 24-h tidal component above 130 km /15/, reducing the effectiveness of the tidal wind shear mechanism at such heights, and the very small probability of an electric field which could produce an Es layer at such heights (i.e. one directed between N and 30° W of N /14/). The absence of Es below 90 km is partly due to the dumping mechanism /16/, whereby Es layers follow the nodes in the tidal wind only down to a heights where the collision frequency becomes too high for the downward component of the WxB drift to match the vertical phase speed. This height is between 90-100 km for typical tidal wind amplitudes at EISCAT /11/. Electric fields with a southward component can move Es layers to lower altitudes but their speed of descent becomes very small below 90 km altitude. Also, chemical processes which result in neutralization of the ions start to become effective around 90 km altitude. The variation of the height distribution through the day, with mostly higher-altitude layers during the day and progressively lower altitude layers during the evening and early morning hours, fits well with the tidal wind-shears (the daytime layers) and the average electric field behaviour (the evening and early morning layers). The slight difference in altitude distribution of the morning and afternoon tidal layers , where Es layers above 110 km are seen more often in the afternoon than in the morning, can be explained either as a modification by the electric field - early in the morning when the wind-shear node is at higher altitudes, above 115 km, the electric field is usually southward and will prevent Es formation at the wind node - or as an effect of the less distinct wind-shear node in the morning.

The seasonal variation , with much more frequent Es in summer than winter, is less easy to explain. As mentioned above, it is not related to the input of metal ions from meteor ablation, and as we see from Figures 4 and 7, it cannot be explained by the seasonal variation in the tides - it is seen even in the nighttime Es which are formed by electric field, not tides, and the tidal nodes are in any case more distinct in winter than summer. Figures 3 ,7 and 8 offer a possible explanation. It is clear that, at times, metallic ions can be and are transported upwards through the E-region. The winds are at

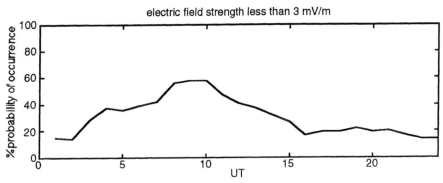

Figure 9. Probability of electric field strength less than 3 mV/m at EISCAT.

times eastward at all heights, and the electric field is often northward. Either condition will result in upward transport of metal ions, usually in dispersed form, but occassionally , if the divergence in vertical ion drift is small, as a distinct layer (Fig. 3, 3rd and possibly 4th example). This raises the possibility that the seasonal variation is due to a reduction in summer of the neutralisation-rate of ions at the lowest altitudes. If fewer ions are neutralised, more are available to be transported upward to form new Es layers.

Another feature which may partly be explained by the wind-shear and electric field mechanisms is multiple layers. Both mechanisms result in a separation of ions with different masses into separate layers due to different gyro and collision frequencies /11,17/. However, the separation is usually less that 1-2 km , and separate layers of the same ions have at times been observed by rockets, so this cannot account for all multiple layers. Another possibility is that multiple layers result from a new layer being transported down to almost join an earlier, 'dumped' layer. This mechanism is beautifully illustrated in /18/. However, this again seems inadequate to explain all multiple layers. A further possibility is that multiple layers are formed by 3-D wind effects /19/ which are not accounted for in the kind of 1-D simulations shown in Figures 5 and 6. It is also possible that underlying layer structures in the neutral atmosphere play a role.

A further unsolved problem is the patchiness and very steep horizontal gradients in Es layers. The final example in Figure 3 is a good illustration of this. The Es layer in this example was detected by ionosonde (with a broad beam) for several minutes both before and after the appearance of the most intense layer in the narrow EISCAT beam. It clearly represents a spatially limited, drifting patch of extremely high density with sharp edges. The uneven time coverage in the example in Figure 3 is because the radar antanna was scanning in a sequence of positions, vertical, to the east and to the south. Only electron density profiles for the vertical position are shown. The information from the other positions shows that the layer was drifting southward at about 40 m/s. The most intense patch is only about 7 km in N-S extent and the layer reduces from full intensity to half intensity in less than 3 km on its southern edge. It remains to be seen if such features can also be explained by 3-D modelling.

A number of other problems also remain. Es layers, when they are measured with good altitude resolution, are most often found to be rather symmetric in altitude, with equally steep vertical gradients on top and underneath. This applies even to layers which appear at low altitude together with southward directed electric fields. This is not in agreement with them having been formed by such electric fields, however, since the simulations show that they should then have much steeper topside gradients than underneath /11/. It is always possible to assume, for individual cases, that they were first formed by a NW directed field, or by a combination of electric field and an upward wind. However, it seems unsatisfactory that this seems to be the rule rather than the exception. It may be that further, as yet unrecognised , processes play a role also here. A possibly related problem is that of the thin neutral metallic layers which are often seen together with Es layers. The known chemical processes which could neutralise ions to produce such layers seem to be insufficient to explain their appearence /20/. Finally, there remains the problem of the slightly thicker thin layers (Figure 1, second example, possibly also Figure 3, 1st example). These seem to be formed by transport of relatively long-lived ions, as are Es layers. However, the winter layers also clearly require solar illumination and disappear rapidly as the sun sets /12/. They cannot therefore be explained by the usual metal ions present in Es ($Fe+$, $Mg+$) since these have lifetimes of several hours (at 100 km).

In conclusion, the broad characteristics of thin ion layers at auroral latitudes can be explained by a combination of tidal wind shear and electric field formation mechanisms. A number of problems remain however, which may be due to effects of 3-D wind fields, and / or to chemical interaction with unexpected minor constituents in the underlying atmosphere.

Acknowledgements The EISCAT Scientific Association is supported by the Centre National de la Recherche Scientifique of France, Suomen Akatemia of Finland, Max Planck Gesellschaft of Germany, Norges Almenvitenskaplige Forskningsråd of Norway, Naturvetenskapliga Forskningsrådet of Sweden and the Science and Engineering Research Council of the United Kingdom. The work of S.K. is supported by the Naturvetenskapliga Forskningsrådet of Sweden.

REFERENCES

1. Turunen, T. Mukunda Rao, M., Sequential Es at Sodankyla, *Geophysica* 13, 175, 1975
2. Turunen, T., Mukunda Rao, M., Statistical behaviour of sporadic-E at Sodankyla 1958-1971, *Geophysica* 14, 77, 1976
3. Turunen T., The diurnal variation of Es layer parameters at Sodankyla in summer 1973 based on ionospheric soundings utilizing low fixed gain, *Geophysica* 14, 55, 1976
4. Turunen, T., The long-term variation of Es layer parameters at Sodankyla 1958-1972, *Geophysica* 14, 61, 1975
5. Turunen, T., Sporadic E layer and magnetic activity at Sodankyla, *Geophysica* 14, 47, 1976
6. Whitehead, J.D., Recent work on mid-latitude and equatorial sporadic-E, *J. Atmos. Terr. Phys.* 51, 401-424, 1989
7. Huuskonen, A., Nygren, T., Jalonen, L., Björnå, N., Hansen, T.L., Brekke, A., Turunen, T., Ion composition in sporadic E layers measured by the EISCAT UHF radar, *J. Geophys. Res.* 93, 14603-14610, 1988
8. Alpers, M., Blix, T., Kirkwood, S., Krankowsky, D., Lübken, F.-J., Lutz, S., von Zahn, U., First simultaneous measurements of neutral and ionized iron densities in the upper mesosphere, *J. Geophys Res.* 98, 275-284, 1993
9. von Zahn, U., Hansen, T.L., Sudden neutral sodium layers : a strong link to sporadic E layers, *J. Atmos. Terr. Phys.* 50, 93-104, 1988
10. Kirkwood, S., von Zahn, U., On the role of auroral electric fields in the formation of low-altitude sporadic-E and sudden sodium layers, *J. Atmos. terr. Phys.* 53, 389-407, 1991
11. Kirkwood, S., von Zahn, U. Formation mechanisms for low-altitude Es and their relationship with neutral Fe layers : results from the METAL campaign, *J. Geophys. Res.* 98, 21549-21561, 1993
12. Kirkwood, S. Anomalous ion layers in the high-latitude winter E-region, *Geophys. Res. Lett.* 18, 1189-1192, 1991
13. Whitehead, J.D., Production and prediction of sporadic E, *Rev. Geophys.* 8, 65-144, 1970
14. Nygrén, T., Jalonen, L., Oksman, J., Turunen, T., The role of electric field and neutral wind direction in the formation of sporadic-E layers, *J. Atmos. Terr. Phys.* 46, 373-381, 1984
15. Brekke, A., Nozawa, S., Sparr, T., Studies of the E-region neutral wind in the quiet auroral ionosphere, *J. Geophys. Res.* 99, 8801, 1994
16. Chimonas, G., Axford, W.I., Vertical movements of temperate zone sporadic-E layers, *J. Geophys. Res.* 73, 111-117,1968
17. Nygrén, T., Jalonen, L., Huuskonen, A, Turunen, T., Density profiles of sporadic-E layers containing two metal ion species, *J. Atmos. Terr. Phys.* 46, 885-893, 1984
18. Turunen, T., Nygrén, T., Huuskonen, A., Nocturnal high-latitude E-region in winter during extremely quiet conditions, *J. Atmos. terr. Phys.* 55, 783-795, 1993
19. Höffner, J., von Zahn, U., 1 and 3-dimensional numerical simulations of the formation of sporadic ion layers below 100 km altitude, *ESA SP-355,* 57-61, 1994
20. Hansen, G., von Zahn, U., Sudden sodium layers in polar latitudes, *J. Atmos. Terr. Phys.* 52, 585-608, 1990

 Pergamon

PII: S0273-1177(96)00047-6

Adv. Space Res. Vol. 19, No. 1, pp. (1)159–(1)168, 1997
© 1997 COSPAR
Printed in Great Britain. All rights reserved
0273–1177/97 $17.00 + 0.00

SMALL-SCALE STRUCTURE OF ELECTRON DENSITY IN THE D REGION DURING ONSET PHASES OF AN AURORAL ABSORPTION SUBSTORM

Hilkka Ranta* and Hisao Yamagishi**

* *Geophysical Observatory, FIN-99600 Sodankyla, Finland*
** *National Institute of Polar Research, 9-10 Kaga 1-Chome, Itabashiku, Tokyo 173, Japan*

ABSTRACT

Small-scale structure of the electron density in the D region during the preceding and onset phases of auroral absorption substorms classified as unloading processes was studied on the bases of imaging riometer measurements made at Tjornes, Iceland, and broad -beam riometer data collected in the Nordic Countries. The precipitation during the onset phase may be localized, covering only 10 km in both north-south and east-west directions. After the substorm onset ionospheric absorption was observed to pulsate with a period of 1 -12 minutes. Possibly this pulsation is related to the magnetohydrodynamic, field line resonances in the nightside magnetosphere and to the near-Earth neutral line substorm model.
©1997 COSPAR. All rights reserved

INTRODUCTION

A magnetospheric substorm is a process initiated on the nightside of the Earth in which a significant amount of energy derived from interaction between the solar wind and the magnetosphere is deposited in the auroral ionosphere and the magnetosphere. As observed from the earth, the onset of a magnetospheric substorm is defined as the time of a sudden brightening of a quiet auroral arc, the rapid onset of a negative magnetic bay and high-energy precipitation /2,5,6,11/. Fundamental to all of these observations is the recognition that a substorm must include at least one auroral breakup followed by a poleward expansion. However, most substorm expansions consist of a sequence of quasiperiodic intensifications of the aurora and electrojet. If one of these is of short duration (about 5 min) and is not followed by a poleward expansion it is referred to as a pseudo-breakup. The first in a sequence of intensifications producing a poleward expansion is referred to as the expansion phase onset. Subsequent activations in the sequence are called intensifications.

A typical sequence of an isolated substorm starts around the midnight meridian with the brightening and activation of a quiet auroral arc, which then expands polewards. The activity also expands eastward and westward and a westward travelling surge often develops. This lasts about half an hour, after which the aurora breaks up into patches and gradually returns to a quiet activity like that preceding the substorm. The above describes an isolated storm. During active periods the discrete phases may become blurred due to ovelapping substorms and to a continuous auroral enhancement that may be of quite a different nature. It has also been shown that a magnetospheric substorm contains a growth phase /4,9/. During such growth phases, satellites in syncronous orbit (6.6 Re) have observed a gradual weakening and deformation of the magnetic field near midnight, suggesting , that its field lines become stretched nightward. The stretched magnetic field store energy, which is later suddenly released in the substorm and transmitted to accelerated particles. Indications suggest , however, that during continuous substorm activity a mode of direct energy transfer from the solar wind to the magnetosphere is enhanced , and that much of the energy then released had not been stored in the tail but directly transmitted /2,3,11,12,13/.

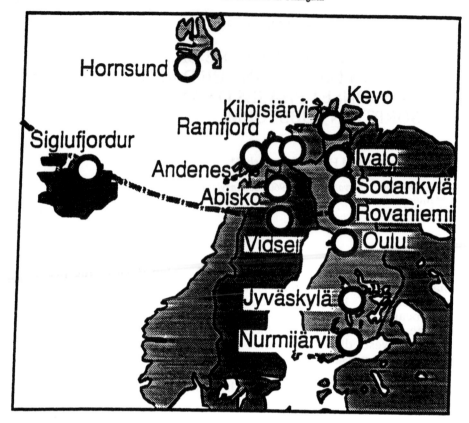

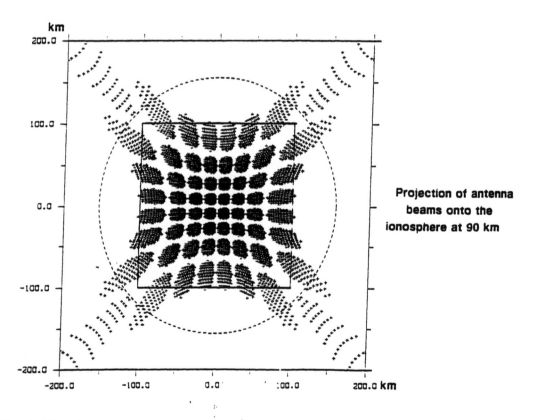

Projection of antenna
beams onto the
ionosphere at 90 km

Fig. 1. The riometer stations in the Nordic countries and the projection of the 3dB
contours of the individual beams at the 90-km level at Tjornes in Iceland.

A magnetic reconnection process in the Earth's magnetosphere was proposed in 1961, primarily as a mean of explaining a connection between field lines of the solar wind and those of the Earth's field. It was proposed that two neutral points were formed: the first on the day side where a connection with the Earth was established and the second on the night side, where it was again broken off. In between these points, polar field lines of the Earth were connected to interplanetary space and able to draw energy from the motion of the solar wind there. Theoretical reasons existed to expect that the process became amplified when the IMF pointed southward and was weakened when it pointed northward, in which case field lines would need to be greatly deformed to make the connection. The discovery of plasma -sheet thinning suggested an alternative explanation: that the tail's plasma sheet became pinched fairly close to Earth forming an additional near-Earth neutral line.

Auroral absorption (AA) substorms caused by ionization by energetic electrons in the D region during a magnetospheric substorm have been intensively studied on the bases of riometer measurements with broad beam antennas /7,8,9,10/. However, since broad beam antennas cover an area of approximately 100 km x 200 km at the height of 90 km and the detailed variation of the electron density in the D region cannot be studied on that basis. In this paper we report a study of the small-scale structure of the D region electron density during AA substorms, based on the measurements collected with the scanning narrow-beam antenna of an imaging riometer. Three of the substorms are examined in detail.

MEASUREMENTS

The time intervals given in Table 1 were selected to study the behaviour of an auroral absorption substorm during the preceding and breakup phase of the substorm.

TABLE 1. The selected time intervals

time			observed by imaging riometer:
6 March 1991	2016- 2112	UT	sharp onset
2 May 1991	1744-1800	UT	sharp onset
24 March 1991	2140-2147	UT	sharp onset
25 June 1991	2001-2207	UT	sharp onset
9 September 1991	2028-2107	UT	sharp onset
10 May 1991	1823-2230	UT	sharp onset
25 June 1992	2002-2107	UT	pred. phase and sharp onset

Two substorm events occurring on 9 September 1991 and 2 May 1991 are discussed in this paper in detail. All riometer measurements made in the Nordic Countries during the selected time intervals were analysed. The stations are located between L-values 3.3 and 13.4 and between longitudes 341° E and 30° E. The operating frequencies vary between 27.6 and 32.4 MHz. A list of the riometer stations with geographic coordinates and operating frequencies is given in Table 2. The location of the stations are shown in Figure 1.

The imaging riometer data were obtained from the station at Tjörnes in Iceland operating at a frequency of 30 MHz. The imaging riometer antenna consists of 64 wire dipoles arranged in 8x 8 array and this forms 64 narrow antenna beams (- 3 dB beam width of 12 degrees), which cover an area of 400 km by 400 km at the altitude of 90 km(see Fig. 1). The image is obtained every 4 seconds /14/.

The variation in the Earth's magnetic field was recorded by the IMAGE magnetometer network in the Nordic Countries including 7 stations between latitudes 77° N and 66° N . The stations are listed in Table 3.

CASE STUDIES OF INDIVIDUAL AURORAL ABSORPTION SUBSTORM

9 September 1991

On 9 September 1991 between 12 and 24 UT, Kp-values were 5, 5-, 5 and 7. A substorm event having several onsets commenced after 1800 UT.

The substorm activity began with ionospheric absorption on 9 September 1991 after 1900 UT in the Finnish longitudinal sector (not shown). The first onset was seen at 1942 UT at Sodankylä and Rovaniemi, the second one at 2015 UT at Kevo, Sodankylä and Rovaniemi and the third one at 2030 UT at Sodankylä and Rovaniemi. The third onset was seen a few minutes later, at 2033 UT, at Siglufjördur, in Iceland. The substorm onset observed at Siglufjördur has features of a multi onset. Increases in absorption were seen at Siglufjördur at 2033 - 2050 UT, 2050 -2100 UT and 2102 -2120 UT.

The first sharp increase of absorption was picked up by the imaging riometer at Tjörnes at 2032 UT in 3 southernmost beams (Fig. 2 and 3) . Five minutes later, at about 2037 UT, a new increase of absorption was seen by all beams, from north to south and from east to west. The imaging riometer data clearly illustrate the localized features of energetic electron precipitation during the onset phase of a substorm. In both north-south and east-west directions the precipitation region may be only 10 km wide. The absorption increase observed by the broad-beam antenna as a single peak of absorption includes pulsation of ionospheric absorption with periods of 1-4 minutes.

During the second increase of absorption, between 2050 and 2100 UT, an absorption increase was observed in the northern- and westernmost beams at 2051.01 UT. Thereafter the precipitation region enlarged westward and southward. The strongest absorption during this increasing period was observed by the easternmost beams. The antenna array was not completely covered by absorption and the structure of the absorption area differed from that seen during the first increase in being smoother character.

2 May 1991

On 2 May 1991, between 12-24 UT, Kp values were 6, 6-, 5 and 4+.

Substorm activity started to develop after 1646 UT. The preceding bay was observed at Rovaniemi, Vidsel and Oulu (not shown). Maximum absorption during the preceding phase, 2.8 dB, was recorded at Oulu . At Vidsel maximum absorption was 2.1 dB, and at Rovaniemi 1.2 dB. In this , as the event in September the pulsation form was most clearly seen at Oulu. The period of pulsation was 10 minutes. Both the movement of the preceding bay southward and intensification during the movement were clearly observed. The onset of the substorm was recorded at Oulu at 1732 UT, at Vidsel at 1736 UT, at Rovaniemi at 1740 UT and at Kilpisjärvi at 1744 UT.

The imaging riometer at Tjörnes picked up the substorm onset with its two southernmost beams at 1746.50 UT (Fig. 4 and 5). The precipitation region subsequently enlarged northward, and between 1748 and 1750 UT the whole antenna area was covered by electron precipitation. Pulsations in absorption values were seen after the onset with the periods of 1 to 2 minutes. No phase shifts were noticed.

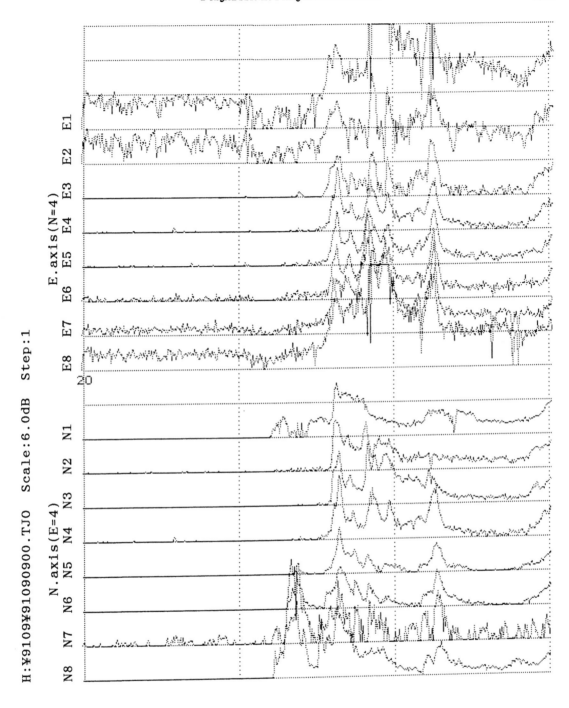

Fig.2. The absorption values measured by the imaging riometer at Tjörnes for eight central beams in west-east and north-south directions on 9 September 1991.

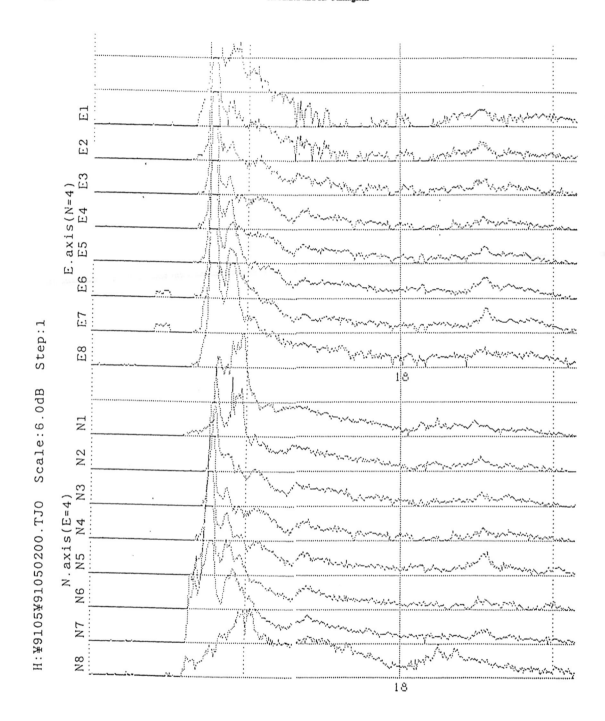

Fig. 4. The absorption values measured by the imaging riometer at Tjornes for eight central beams in west-east and north-south directions on 2 May 1991.

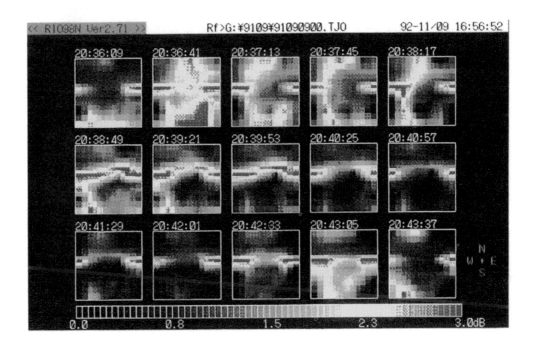

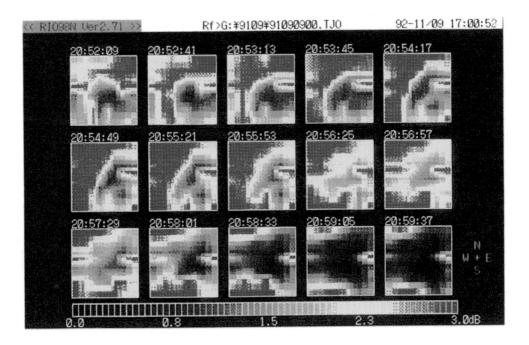

Fig. 3. The absorption values measured by the imaging riometer at Tjörnes on 9 September 1991.

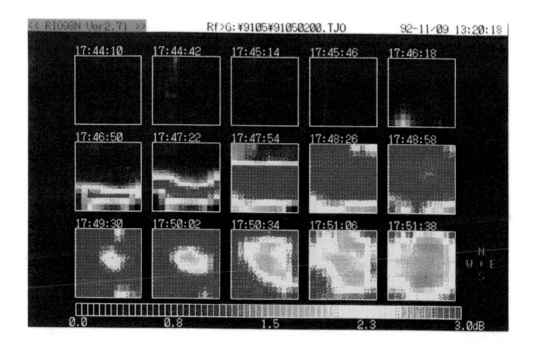

Fig. 5. The absorption values measured by the imaging riometer at Tjornes on 2 May 1991.

DISCUSSION AND CONCLUSIONS

In his original description of auroral substorms Akasofu (1964) characterized the expansion phase as an explosive phenomenon intrinsic to the magnetosphere. However, 15 years later he altered his position, describing the magnetosphere as being primarily driven by the solar wind dynamo /2/. In this later interpretation the expansion phase is simply a nonlinear effect of increasing current along field lines connecting the ionosphere to the solar wind. The existence of a growth phase, however, had led McPherron (1970) to conclude that the energy released in the explosive expansion phase had been stored earlier. To characterize the difference between these two views, Akasofu (1977) introduced the terms driven process and unloading process. A driven process is one for which the waveforms of the solar wind energy input and output parameters are virtually identical, with perhaps at most a small time delay . In contrast, an unloading process is one in which the input wave form is stored without a corresponding output, until after a long time delay the energy is released. As envisioned by Akasofu, unloading is internal to the magnetosphere and hence its wave form bears little resemblance to the wave form of the input. Akasofu (1977) noted the possibility that the magnetosphere actually responds as a superposition of the two types of processes. However, in numerous reports he argued that the magnetosphere is primarily driven .

The substorms studied in this paper were selected on the basis that the onset of the substorm was very sharp. All cases can thus be classified as unloading processes. Based on all the data, including that in Table 1, the following conclusions can be made:
1. In every case a preceding phase was seen before the substorm onset, even during the very disturbed period when Kp was high. During these very disturbed periods the preceding bay was seen at the southernmost stations, at L-values of about 4. During quiet periods the preceding phase was seen at an L-value as high as 13.

2. The preceding absorption bay had a pulsating form with a period of 10 - 20 minutes.

3. A minimum between the preceding and onset phases, the so called fading phenomenon , was not seen at the northernmost latitudes of the region where the substorm occured , but it was seen at the southernmost latitudes .

4.During the onset phase of the AA substorm the precipitation region may be localized and may cover only about 10 km in east-west and in north-south directions.

5. As observed by the imaging riometer, the propagation speed of the substorm onset from south to north was about 1-3 km/s. The propagation speed from east to west was in some cases so rapid that it could not be recorded by the imaging riometer.

6. Pulsation of ionospheric absorption with periods of 1-12 minutes was observed after the substorm onset.

There are probably at least four different modes of substorm. The first is the northward IMF. Many studies indicate that during such conditions reconnection switches from the subsolar point to points poleward of the polar cusps. Strong currents flow in the polar cap, but they are invisible to the stations monitoring the electrojet indices.   The second mode is the one described above which applies to isolated substorms. The third may be relevant to pseudo-breakups, and the fourth to convection bays.

McPherron and Baker (1993) have proposed that a pseudobreakup is the ionospheric manifestation of the near-Earth neutral line   model when reconnection does not succeed in cutting the last closed field lines connected to the distant x-line at the boundary of the plasma sheet. Arguing against this   observations made by Samson et al. (1993) during substorm intensifications indicate that at least one component in the substorm process must be active on dipole -like field lines near the Earth. Magnetometer, riometer   and HF radar data have indicated the presence of magnetohydrodynamic field line resonances in the nightside magnetosphere. These resonances have frequencies of about 1.3, 1.9, 2.6 and 3.4 mHz and are due to cavity modes or waveguide modes, which form between the magnetopause and the turning point on dipole -like magnetic shells. All these observations provide strong evidence, that at least one component of the substorm mechanism must be active very close to the Earth, probably on dipole -like field lines in regions with trapped and quasi-trapped energetic particles. The pulsation of ionospheric absorption   after the substorm onset as observed in this study, supports the model of a near-Earth neutral line as a active mode of a substorm.

REFERENCES

1. Akasofu, S.I. (1964) Planet. Space Sci. 12, 273.

2. Akasofu, S-I. (1977) Physics of Magnetospheric Substorms, Reidel Dordrecht, The Netherlands.

3. Hones E.W. ed. (1984) Magnetic Reconnection in Space and Laboratory Plasma. Geophysical Monographs 30, AGU, Washington DC.

4. McPherron R.L. (1970) J. Geophys. Res. 75, 5992.

5. McPherron R.L. (1979) Rev. of Geophys. and Space Physics. 17,4, 657.

6. McPherron, R.L., D.N. Baker (1993) J. atm. terr. Phys. 55,8, 1091.

7. Pikkarainen, T., Kangas. J., Ranta H., Ranta A., Maltseva, N., Troitskaya, V., Afanasieva, L. (1986) J. atm. terr. Phys. 48,6,585.

8. Ranta, H. (1978) J. Geophys. Res. 83, 3893.

9. Ranta, H., Ranta A., Collis, P.N., Hargreaves, J.K. (1981) Planet. Space Sci. 29,12,1287.

10. Ranta, A., Ranta H., Rosenberg, T.J., Wedeken, U., Stauning P. (1983) Planet. Space Sci. 31,12,1415.

11. Rostoker G., Akasofu S.J., Foster J., Greenwald. R.A., Kamide Y., Kawasaki K., Lui A.T.Y. McPherron R.L., Russell C.T. (1980) J. Geophys. Res. A4, 1663.

12. Samson, L.C., D.D. Wallis, T.J. Hughes, F. Creutzberg, J.M. Ruohoniemi, R.A. Greenwald (1992), J. Geophys. Res. 97, 8495.

13. Stern D.P. (1989) Rev. Geophys. 27,103.

14. Yamagishi, H., T. Kiruchi, S. Ikeba, T. Yoshino (1989) Proc. of the NIPR symposium on Upper Atmosphere Physics 2,100.

TABLE 2. Riometer stations.

Station		frequency MHz	geogr. lat.	geogr. long.
Hornsund	HOR	30.0	$77^{\circ}00'$ N	$15^{\circ}36'$ E
Kevo	KEV	27.6	69 45	27 01
Kilpisjärvi	KIL	30.0	60 03	20 47
Ivalo	IVA	27.6	68 36	27 26
Sodankylä	SOD	27.6	67 25	26 24
Rovaniemi	ROV	32.4	66 34	26 01
Oulu	OUL	27.6	65 06	25 59
Jyväskylä	JYV	32.4	62 24	25 22
Nurmijärvi	NUR	27.6	60 31	24 39
Abisko	ABI	30.0	68 24	18 54
Vidsel	VID	30.0	65 48	20 36
Siglufjördur	SIG	30.0	66 12	341 06
Tjörnes	TJÖ	30.0	66 22	342 50

TABLE 3. IMAGE magnetometer stations

		geogr. latitude	longitude
Söröya	SOR	70.54° N	22.22° E
Masi	MAS	69.46	23.70
Muonio	MUO	68.02	23.53
Pello	PEL	66.90	24.08
Kilpisjärvi	KIL	69.02	20.79
Kevo	KEV	69.76	27.01
Oulujärvi	OUL	64.52	27.23
Hankasalmi	HAN	62.30	26.65
Nurmijärvi	NUR	60.50	24.65

Adv. Space Res. Vol. 19, No. 1, pp. (1)169–(1)173, 1997
© 1997 COSPAR
Printed in Great Britain. All rights reserved
0273–1177/97 $17.00 + 0.00

Pergamon

PII: S0273-1177(96)00048-8

THIN LAYERS OF IONIZATION OBSERVED BY ROCKETBORNE PROBES IN EQUATORIAL E REGION

S. P. Gupta

Physical Research Laboratory, Navrangpura, Ahmedabad 380 009, India

ABSTRACT

Rocketborne experiments were conducted from two Indian Rocket Ranges, namely Thumba and Sriharikota (SHAR) situated at dip = 0 and dip = 12° respectively. Electron density, ion density and electric field were measured in day and night. On some occasions near simultaneous measurements of neutral winds were also carried out in altitude region 90-120 km. During night time, the thin layers of ionization with electron density 3 to 10 times of ambient density were observed over SHAR as well as Thumba. These layers have thickness of 1 km to 5 km and have separation of 5 km to 10 km. Over SHAR the layers were much denser compared to Thumba. Occasionally such layers were seen in day time also. We suggest that gravity waves and tidal wind waves are main cause of these layers.
©1997 COSPAR. All rights reserved

INTRODUCTION

In the equatorial E region, several types of irregularities are present with scale size 10 km to a few meters. These irregularities have life time of an hour to about a day. The meter scale size irregularities are produced due to plasma processes like ExB plasma instability or two stream (associated with electrojet) instability, where E is vertical polarization field and B is earth's magnetic field which is horizontal over equator and directed towasrds north. The generation mechanism of these small scale irregularities is more or less understood atleast in a broad way /1,2/. How km (in vertical direction) scale size irregularities are generated over equatorial latitude in 100 km region is still not well understood. One of the mechanisms for generation of km scale size irregularities can be wind shear mechanism /3/. Over magnetic equator wind shear mechanism may not be operative since the vertical polarization electric fields are stronger than wind induced fields. However, above 110 km where vertical electric field becomes small, it may be possible that wind shear mechanism may work. Away from the equator at SHAR which is about 6° away the wind shear mechanism may work. Assuming that the polalrization field intensity at SHAR becomes very small.

Rocketborne studies of km to 1 meter scale size irregularities have been carried out over Thumba (dip = 0) and SHAR (12°) during different epoch of solar activity. We still mostly consider only night time results. Electron density, ion density and electric field and on some occasions neutral winds were also measured near simultaneously. The groundbased experiment like ionosonde and three station drift techniques were operating at the time of rocket experiments /4/. We also want to mention that the vertical electric field in equatorial E region 90-140 km is upward in day and downward in night. The reversal takes place at sunrise and sunset /5/.

RESULTS

Rocketborne Langmuir technique was used from both the locations, i.e. Thumba and SHAR to investigate electron density irregularities. The technique was identical in both the cases. Hence we can compare the results. The sensor is a spherical sensor with guard electrode. A fixed bias of +4 volt with respect to rocket body is applied. However, in several flights, a sweep voltage from -4 volt to +4 volt was applied. But here we discuss only the results where the sensor was operated at fixed bias.

Figure 1 shows the results from Thumba where we see irregularities of wave type nature with vertical wavelengths of the order of 5 km. This was in the month of March. In earlier occasion we had conducted three rocket flights on same night in March 1967. We can see that we do not see such structures (Figure 2). Only in morning flight and evening flight there is layer at 115 km and 130 km respectively. On several occasions, sometimes we see these irregularities and sometime we do not see /6/. This shows that the mechanism of producing these structures is not diurnal winds, but some other mechanism which has life time of the order of an hour. Such mechanism can be gravity waves.

Now we show the results from SHAR for the period February-March 1982. Figure 3 shows the results of evening flight conducted on 16 February and 1st March, 1982 at around same time. We can see that on both occasions there are layers in 90-120 km regions. Figure 4 shows the results for day and night flights from SHAR on different days. We can see that even at day as well as in night the layers are present. Over SHAR the thin layers are seen most of the time during night. While over Thumba, the layers are seen only occasionally. We are considering only night time cases.

DISCUSSION

Our results show that thin layers in equatorial E region can be produced by two different mecahnisms, namely (i) gravity waves operating at magnetic equator Thumba (dip = 0) and (ii) Tidal (diurnal tide) through windshear mechanism operates at SHAR (dip = 12°). Wind shear mechanism is not very effective over magnetic equator as discussed earlier. We have conducted simultaneous measurements of neutral winds both from Thumba and SHAR and based on wind measurements we found that wind shear plays vital role in producing layers over SHAR, but over Thumba the wind shear mechanism is not effective /7/. We also want to mention that layer density over SHAR is about ten times more than layer density over Thumba during night time. This also shows that different mechanisms are operative at these locations. When we compare our results with mid-latitude station like Wallops Island reported by Smith /8/, we find that SHAR results closely resemble with Wallops Island results. SHAR is located at the edge of the electrojet region.

Figure 5 shows the spectral index plot. The X axis is 1/L where L is vertical scale size in km. The Y axis is square of percentage amplitudfe, i.e. energy associated with a particular scale. The scale size 100 m to 10 km scale follows a power law with a slope of -2.5 which is close to standard value /9/. The spectral index -3 which is in agreement for 1 km size structure as shown by Weinstock /10/.

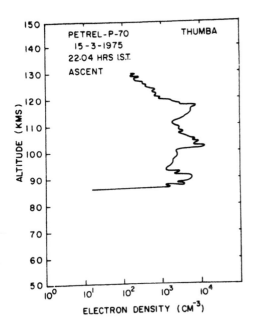

Fig. 1. Electron density profiles from Thumba during night time.

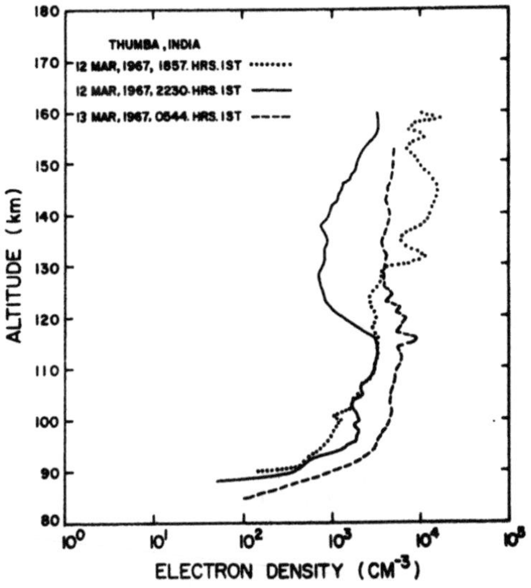

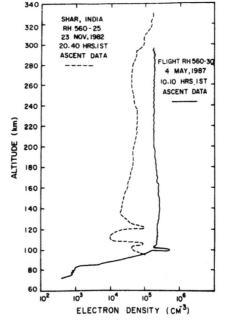

Fig. 2. Electron density profiles from Thumba at different times of night,

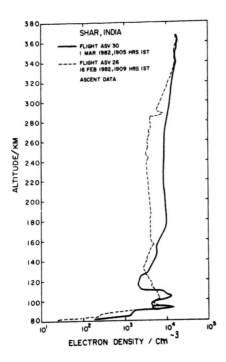

Fig. 3. Electron density profiles from SHAR on different days during evening time.

Fig. 4. Day and night time electron density profiles from SHAR.

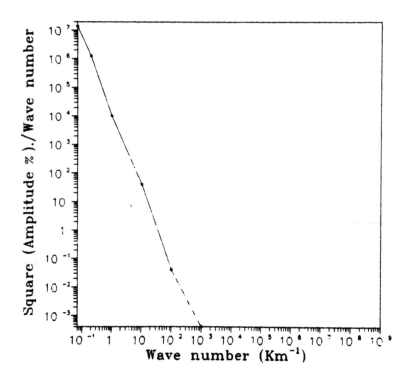

Fig. 5. Wave number spectrum of electron density fluctuations for a night tine rocket flight fron Thunba at 100 kn. altitude.

ACKNOWLEDGEMENTS

The project was funded by the Department of Space, Government of India.

REFERENCES

1. Satya Prakash, S.P. Gupta, B.H. Subbaraya and C.L. Jain, Electrostatic plasma instabilities in the equatorial electrojet, NATURE, Physical Sciences, 233, 56-58 (1971).

2. Satya Prakash, S.P. Gupta and B.H. Subbaraya, Cross field instability and ionization irregularities in the equatorial E region, Nature, Physical Sciences, 230, 170 (1971).

3. W.I. Axford, The wind shear theory of the formation of temperature zone sporadic-E layer, Space Research VII, 126 (1967).

G. Chimonas and W.I. Axford, Vertical movement of temperate zone sporadic E layer, J. Geophys. Res., 73, 111 (1967).

J.D. Whitehead, Difficulty associated with wind shear theory of sporadic E layer, J. Geophys. Res., 76, 3127 (1971).

4. H. Chandra and G.D. Vyas, Equatorial electrojet and ionosphere drifts at low latitudes, Geological south of India, Memoir # 24, 389 (1992).

5. S.P. Gupta, Formation of sporadic-E layer at low magnetic latitudes, Planet. & Space Sci., 34, 1081 (1986).

S.P. Gupta and H. Chandra, Role of vertical electric field and neutral winds in the formation of blanketing type of sporadic E layer over Thumba, Geological Society of India, Memoir # 24, 401 (1992).

6. S.P. Gupta, Ionization layers over the magnetic equator during meteor shower days, Adv. in Space Res., 10, 105 (1990).

7. R. Sridharan, R. Raghavarao, R. Suhasini, R. Naryanan, R. Sekar, V.V. Babu and V. Sudhakar, Winds, wind shear and plasma density during initial phase of a magnetic storm from equatorial latitude, J. Atmos. and Terr. Phys., 51, 169 (1989).

8. L.G. Smith, Rocket observations of sporadic-E and related features of E region, Radio Sci., 1, 178 (1966).

9. E.M. Dewan and R.E. Good, Saturation and universal spectrum for vertical profiles of horizontal scalar winds in the atmosphere, J. Geophys. Res., 91, 2742 (1986).

10. J. Weinstock, Theoretical gravity wave, Spectrum, Radio Sci., 20, 1295 (1985).

APPENDIX

Adv. Space Res. Vol. 19, No. 1, pp. (1)177–(1)180, 1997
© 1997 COSPAR
Printed in Great Britain. All rights reserved
0273–1177/97 $17.00 + 0.00

Pergamon

PII: S0273-1177(96)00010-5

TOTAL IONOSPHERIC ELECTRON CONTENT OBSERVATION AND ETS-V BEACON EXPERIMENT FOR STUDYING GHz-BAND SCINTILLATION IN THE EQUATORIAL ZONE

K. Igarashi,* M. Nagayama,* A. Ohtani,* N. Hamamoto,*
Y. Hashimoto,* T. Ide,* H. Wakana,** T. Ikegami,**
S. Taira,** S. Yamamoto,** E. Morikawa,** K. Tanaka,**
Utoro Sastrokusumo,*** M. W. Sutopo,*** Narong
Hemmakorn,† Apinan Manyanon,† G. H. Bryant‡ and Kevin
Maitava[1]

* Communications Research Laboratory, Koganei-shi, Tokyo 184, Japan
** Kashima Space Research Center, Communications Research Laboratory,
Kashima-machi 314, Japan
*** Institute of Technology Bandung, Jl Ganesha 10, Bandung 40132,
Indonesia
† King Monkut's Institute of Technology, Chaokuntaharn Ladkrabang,
Bankok 10520, Thailand
‡ The Papua New Guinea University of Technology, Lae, Papua New Guinea
[1] The University of South Pacific, Suva, Fiji

ABSTRACT

A L–band CW–beacon experiment using ETS–V satellite for mobile satellite communications experiments was made with the small satellite ground stations installed at Fiji, Indonesia, Papua New Guinea, Thailand and Japan in order to study propagation effects on satellite–to–ground–links in the equatorial zone. GPS observations of total ionospheric electron content and monitoring of geomagnetic variations were started at the same ground stations in Bandung, Indonesia and Bangkok, Thailand from November, 1993. During the severe ionospheric disturbances on February 22, 1994 Japanese Broadcasting Satellite BS–3a stopped its broadcasting for about one hour by an unexpected satellite trouble. We focused on this event by comparing between TEC observed in low–latitude and mid–latitude of northern hemisphere. The cause of this sattelite trouble is supposed mainly by the electric static discharge (ESD) in the satellite. The severe geomagnetic storm on 21 February, 1994 is related in this paper as one of the cause for this satellite trouble.
©1997 COSPAR. All rights reserved

GPS OBSERVATIONS OF IONOSPHERIC TOTAL ELECTRON CONTENT

Figure 1 shows the map of ground stations for the earth–space propagation experiments in the equatorial zone (ESPREE). The first propagation experimental results using ETS–V satellite of very severe ionospheric scintillation over 20 dB peak–to–peak were reported by Hamamoto et al. /1, 2/. In order to compare the ionospheric response between the low–latitude and the mid–latitude during severe scintillations and magnetic storms GPS observations of total ionospheric electron content and monitoring of geomagnetic variations are carried out in Bandung (6.5° S, 107.3° E geographic; 17.9° S

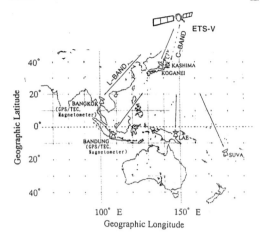

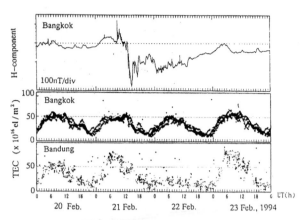

Fig. 1. Ground stations of PARTNERS project
for the earth–satellite propagation experiments
in the equatorial zone (ESPREE).

Fig. 2. TEC data at Bangkok and Bandung for
February 20– 23, 1994. The magnetic field
data at Bangkok are shown in upper panel.

geomagnetic latitude) and Bangkok (13.5° N, 100.3° E geographic; 2.8° N geomagnetic latitude). The ionospheric total electron content measurement using global positioning system is very conienient method to monitor the ionospheric behavior continuously /3/. Figure 2 shows the results of ionospheric TEC observation with GPS receiver (NITSUKI Model 7633) and magnetogram in Bangkok around the SSC at 09h01m UT on 21 February, 1994. From 1h to 2h LT (from 18 h to 19 h UT) in Bangkok night time enhancements in ionospheric electron content appeared clearly every night except 22 February, 1994 /4/. The TEC in Bangkok seems to decrease after 6 h UT on 22 February, comparing with the TEC before the SSC event. The TEC in Bankok increased a little on 23 February /5/. On the other hand the TEC in Bandung decreased to about − 60 % below the normally TEC variation level on 22 February /6, 7/. The TEC of Bangkok and Bandung on 23 February increased more than the TEC on the SSC event. Figure 3 shows the TEC contours in order to see latitudinal variations of this TEC storm behavior. The structure of TEC variation in Figure 3 looks complicate, but TEC depletion region on February 22 was very wide in latitudinal directions. On 23 February the region of TEC enhancement extends to the latitudinal directions. There is not so much depletion of TEC in Bangkok on 22 February, 1994.   The TEC of north side from Bangkok decreases about 10×10^{17} el / m^2. High TEC region in Bangkok expandes to south direction, following on 22 February, 1994.

A COMPARISON OF EQUATORIAL ZONE AND NORTHERN HEMISPHERE TEC

Figure 4 shows the TEC observed in Kokubunji (35.7° N, 139.48° E geographic; 25.5° N, 205.8° E geomagnetic) , Yamagawa (31.2° N, 130.6° E geographic; 20.4° N, 198.3° E geomagnetic) and Okinawa (26.3° N, 127.8° E geographic; 15.3° N, 196.0° E geomagnetic). Faraday rotation observations were conducted at Yamagawa and Okinawa utilizing beacon transmission from a geostationary satellite GOES–3. The TEC observation at Kokubunji was made with the same GPS receiver as PARTNERS project. Before SSC at 9h01m UT on February 21 the TEC enhancement appears in Okinawa and Yamagawa at about 13h JST. Just after the SSC a small TEC enhancement was observed in Yamagawa, Okinawa and Bangkok clearly /8, 9/. Nighttime enhancement of TEC in

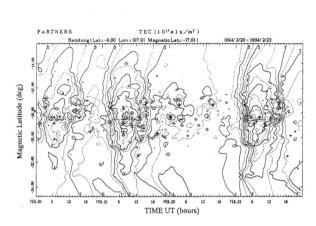

Fig. 3. TEC contours of Bandung for February
20 – 23, 1994, versus geomagnetic latitude and
universal time. The contours give ionospheric-
TEC values in units of 10^{17} el / m^2.

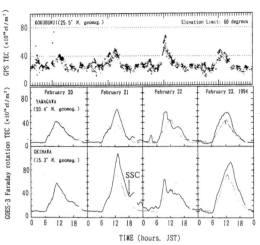

Fig. 4. TEC variations for February 20 – 23,
1994 at Kokubunji, Yamagawa and Okinawa.
Dashed line shows the monthly median of TEC.

Bangkok seems to be correlated with the nighttime TEC enhancement in Yamagawa and Okinawa. The TEC in Okinawa decreased about 30 % of monthly median TEC value during the severe TEC depletion event observed in Bandung on 22 February. This TEC depletion appeared in the latitude of Yamagawa and Kokubunji too. Negative storm effects are caused primarily by composition changes /8/. Around 9 h JST (JST=UT + 9 hours) on 22 February TEC enhancement appeared in Okinawa, Yamagawa and Kokubunji at the same time. Single-day enhancements of TEC are often observed in mid-latitude. This is generally due to magnetic sudden commencement in the winter /10/. During this event Japanese broadcasting satellite BS-3a stopped the transmission during a live broadcast of the Winter Olympic Game at about 21h 14m JST (12h 14m UT) on 22 February, 1994. Figure 5 shows the field strength of BS-3a in 12 GHz band at Yamagawa of CRL. The particle data from SEM (Space Environment Monitor) of Japanese Meteorogical Satellite GMS-4 is shown in Figure 6. At 03 h00m UT on 20 February, 1994 a proton event more than 10 MeV was observed. This event was biggest proton event since March in 1993. Just before 12 h UT on February 22, 1994 there seems to be a rapid increase of

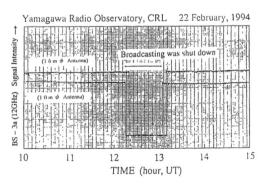

Fig. 5. The signal intensity record of the Japanese
broadcasting satellite BS-3a in 12 GHz band
when the broadcasting was shut down on
February 22, 1994 by an unexpected trouble.

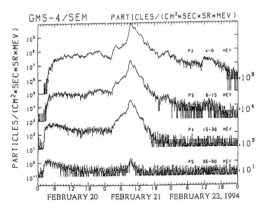

Fig. 6. Proton flux data obtained with the SEM
of GMS-4 from 20 to 22 February, 1994

proton flux level in 4–8 MEV range. The cause of this satellite trouble was explained as the electro-static discharge (ESD) effect by the National Space and Developing Agency (NASDA) in Japan. Because the satellite was in the Harang discontinuity where a shear exists in the flow of plasma.

CONCLUSIONS

Multistational L–band beacon experiment by using the satellite ETS–V was began in the equatorial zone as one of the PARTNERS projects. We have started TEC and magnetomer observations at the same sites of small satelllite ground stations in Bandung and Bangkok from November in 1993. The TEC behavior associated with the SSC on 21 February, 1994 was presented. In Bandung TEC depletion appeared during the recovery phase of the magnetic storm. In Bangkok remarkable TEC depletion on 22 February, 1994 was not appeared. Nighttime enhancement of TEC was observed at about 1 h LT in Bangkok. The TEC behaivior in the northern mid–latitude station showed different features, especially on 22 February, 1994. During this event an unexpected trouble of Japanese broadcasting satellite BS–3a becomes one of the important study subjects in the space environment change effects on geostationary satellite.

ACKNOWLEDGMENTS

This experiment was supported by the PARTNERS promotion council which is an organization made up mainly the Ministry of Posts and Telecommunications (MPT) and the National Space Development Agency (NASDA) in Japan. We wish to acknowledge the Space Policy Office, MPT for their support and encouragement. We are grateful to Kiyoshi Sato, JAICA expert, KMITL for his help in making continuous observation in Thailand. We also wish to thank K. Kondo and R. Suzuki of National Institute of Multimedia Education for their helpful discussion. The SEM data of GMS-4 was provided by the Japan Meteorological Agency. We appreciate the cooperation of T. Sulaeman and H. Sulaiman of the National Institute of Aeronautics and Space in providing the Indonesian ionosonde data.

REFERENCES

1. N. Hamamoto, T. Ida, M. Nakata, Y. Kobayashi, R. Miyauchi, and T. Ohshima, 1993 Asia–Pacific Conference on Communications, 1A.1, Taejun, Aug. (1993)

2. I. Ide, Y. Hashimoto, H. Hamamoto, T. Ikegami, K. Tanaka, S. Tokunaga and T. Uekawa, B–178, 1993 IEICE Conference, Sep. (1993)

3. Gabor E. Lanyi and T. Roth, *Radio Science,* 23 (4), 483–492 (1988)

4. N. Balan, G. J. Bailey, R. Balachandran Nair, and J.E. Titheridge, *J. Atmos. Terr. Phys.*, 56 (1), 67–79 (1994)

5. R.S. Dabas, P.K. Bhuyan and T.R. Tyagi, *Radio Science,* 19 (3), 749–756 (1984)

6. M. Mendillo. and J. Baumgardner, *J. Geophys. Res.,* 87 (A9), 7641–7652 (1982)

7. E.J. Weber, H.C. Brinton, J.Buchau and J.G. Moore, *J. Geophys. Res.,* 87 (A12), 10,503–10,513 (1982)

8. J.E. Titheridge and M. J. Buonsanto, *J. Atmos. Terr. Phys.*, 50 (9), 763–780 (1988)

9. M. Mallis and E. A. Essex, *J. Atmos. Terr. Phys.*, 55(7), 1021–1037 (1993)

10. H. Soicher and F. J.Gorman, *Radio Science,* 20 (3), 383–387 (1985)

 Pergamon

Adv. Space Res. Vol. 19, No. 1, pp. (1)181–(1)186, 1997
© 1997 COSPAR
Printed in Great Britain. All rights reserved
0273–1177/97 $17.00 + 0.00

PII: S0273-1177(96)00012-9

THE SHAPES OF THE ASTEROIDS

P. Farinella* and V. Zappalà**

* Dipartimento di Matematica, Università di Pisa, Via Buonarroti 2, 56127

Pisa, Italy

** Osservatorio Astronomico di Torino, 10025 Pino Torinese (Torino), Italy

ABSTRACT

The known asteroid population spans some five orders of magnitude in sizes, up to a diameter of about 900 km. As a consequence, it happens to include the transition from the shape regime dominated by material strength, to that controlled by self-gravitational forces. Most small asteroids are probably irregular rocky boulders, generated by catastrophic fragmentation events, while the largest ones would instead look like regularly-shaped small planets. However, there are probably exceptions to this rule, due to variations in self-gravity, material properties and collisional history. In the intermediate transition range, in particular, a variety of shapes and surface morphologies are probably present.
©1997 COSPAR. All rights reserved

INTRODUCTION

At sizes of a few hundreds km, the transition occurs between celestial bodies whose global shape can be defined as *irregular*, namely can keep for long times a memory of their origin and/or history, and bodies for which self-gravity is intense enough to overcome the solid–state material rigidity, so that — if one neglects small–scale surface topography — their shapes are relaxed to equilibrium figures, moulded exclusively by gravitational, rotational and (possibly) tidal forces (see /1/, Sec. 14.7). The transition size range is very interesting, as there one can observe a variety of peculiar morphologies, providing information on the accretion and destruction processes which took (and take) place in the Solar System. This study has actually progressed rapidly since the 70s, when space probes have started to send to Earth a wealth of close–up images of small planetary satellites, clearly showing the dominance of irregular shapes for diameters smaller than about 200 km, of nearly–spherical equilibrium shapes beyond about 500 km, and a complex transition regime in between /2,3/.

While spanning the same size range, the asteroid population is much more numerous than that of small satellites, and has a different (dominantly rocky instead of icy) composition. Although to date only one of these bodies has been approached by a space probe, Earth-bound observational techniques /4,5/ have provided planetary astronomers with a fairly detailed view of their global shapes and morphologies as a function of size. In the remainder of this paper, we are going to summarize this view, along with some of its implications for the origin and history of the asteroid belt.

LARGE ASTEROIDS

Lightcurve amplitudes, defined as the peak–to–peak variations of the asteroid brightness over a rotational cycle, provide a crude indicator for asteroid shapes. This is true in particular when lightcurves taken at several different oppositions (i.e., viewing orientations) are available: the amplitude then varies from zero, when the body is viewed pole–on and no change in the projected surface area is seen, to a maximum value, which corresponds to the equatorial view and gives an estimate of the global deviation of the shape from an axisymmetric one. For a triaxial ellipsoid of uniform brightness, spinning around the shortest axis, the maximum amplitude (in magnitudes) would just be given by 2.5 times the logarithm of the ratio between the two equatorial axes. When many lightcurves are available, a number of methods have been developed to derive at the same time best–fitting estimates of the global shape of the body and of its polar axis direction /5/.

Most asteroids larger than 150 km in diameter have lightcurves amplitudes not greater than 0.2 mag; the average amplitude is definitely lower than for smaller bodies, and suggests that self-gravitational forces are important in moulding their global shapes /6/. Quite often, moreover, the lightcurves of these large asteroids show irregular patterns of maxima and mimima, whose number is sometimes larger than two per cycle. In some oppositions, twice the number of extrema observed during other oppositions are present, causing ambiguities in the determination of the rotational period. A likely explanation of these phenomena is that the global shape has relaxed to match one of the axisymmetric Maclaurin ellipsoidal figures, which are the stable equilibrium figures of homogeneous spinning bodies when their angular momentum of rotation is small enough /7/, but at the same time the surface is covered by large–scale albedo variegation features, causing brightness variations of the order of 0.1 mag.

A typical such case is probably that of 4 Vesta, the brightest and third largest asteroid. Two independent observational techniques — lightcurve photometry and speckle interferometry /8/ — have shown that a significant, hemispheric–scale albedo contrast is present on its quasi–equilibrium surface, formed by a basaltic crust of igneous origin that has probably undergone giant impact cratering events. The orientations of the rotation axis derived by the two techniques are also close, as well as the polar flattening values. This latter quantity can be used, together with the known rotational period, to constrain the average density of Vesta, which comes out to be in the range 2.4 – 3 g/cm^3. Until a few years ago, this result was at least $\approx$ 30% lower than the density derived by the available values of the mass and volume of the asteroid; but recent occultation data /9/ have shown that Vesta's diameter is about 560 km, some 10% larger than it was assumed earlier, thus removing any discrepancy with the mean density derived under the assumption of an equilibrium shape.

Stellar occultations, albeit occurring sporadically, have in general the potential of yielding very accurate sizes and limb profiles of the involved asteroids. The results obtained so far have been reviewed in /10/. A puzzling case is that f 2 Pallas, for which two different occultations have been observed; coupling these data with lightcurve–derived information on Pallas' pole and rotational phase, a failry smooth but globally triaxial shape has been derived /11/. Since Pallas has about the same size as Vesta, if this result will be confirmed it may point to interesting differences in the thermal and/or collisional histories of the two bodies: the primitive (Pallas) vs. differentiated (Vesta) structure and composition may have led to different strengths and morphologies of the surface layers, which have then undergone different collisional fluxes at different average impact velocities (resulting from Pallas' unusually inclined and eccentric orbit).

JACOBY ELLIPSOIDS AND PILES OF RUBBLE

Although most large asteroids have fairly low lightcurve amplitudes, in the diameter range between 150 and 300 km a group exists of Large Angular Momentum Asteroids (LAMAs), with comparatively large amplitudes (i.e., strongly triaxial shapes and large moments of inertia) and short rotation periods /12,13/. As shown in /6/, if one considers all the asteroids larger than 125 km diameter, a clear correlation is present between amplitude and rotation rate. The upturn towards higher amplitudes occurs at spin rates between 4 and 6 rotations/day, precisely in the range where triaxial (Jacobi) ellipsoids are possible equipotential figures for self–gravitating spinning bodies with negligible strength and constant density, provided the latter quantity is in the range between 1.5 and 3 g/cm^3 /7/. These bodies may be gravitationally bound rubble piles, i.e., asteroids composed of fragments which are held together by their mutual gravity: in turn, this may be a common outcome of catastrophic collisions shattering the target body and transferring to it a large amount of angular momentum of rotation, without ejecting most fragments at speeds exceeding their mutual escape velocity. When the transferred angular momentum exceeds the rotational fission threshold, the same mechanism may give rise to a nearly–contact binary system, which would display a normal rotation period (say, between 6 and 12 hours) but a high amplitude, due to the mutual eclipses of the components /14/.

Extensive observational programs aimed at deriving rotational periods, poles and shapes of most large asteroids have been carried out in the 80s /15/. Although lightcurves do not provide shape determinations accurate enough to really test the hypothesis that any specific body is a rubble pile and matches a hydrostatic equilibrium shape, several asteroids have been identified which are good rubble–pile triaxial or binary candidates, with reasonable mean densities implied by the equilibrium hypothesis. Of course, a rubble pile would not have a very smooth and regular surface, unless much of its mass has been mobilized and finely fragmented by impacts. For the typical size distribution of fragments generated by high–velocity impacts, the mass fraction contributed by bodies of size smaller than 1% of the largest fragments would be of the order of 10%, i.e. substantial but possibly not sufficient to smooth out the large–scale bulges or depressions related to the irregular shapes of the largest reaccumulated objects.

SMALL ASTEROIDS

With the exception of Gaspra, which we are going to discuss separately, the available information on the shapes of the small asteroid population (say, under 50 km diameter) is of a statistical nature. For every specific object, it is rare that lightcurves have been derived at different oppositions, hence the aspect angle is unknown and the lightcurve amplitude is in general smaller than the maximum value, corresponding to the equator–on view. In general, it is assumed that the polar axes are distributed isotropically on the celestial sphere, and in this case the mean value of the aspect angles observed at one opposition will be 60°. Fragments derived from laboratory impact fragmentation experiments can then be treated in the same way, assuming that they are observed from random directions and deriving the corresponding lightcurve amplitude distribution: such comparisons, carried out in the 80s with more and more abundant samples of data, have shown a good agreement between the average global shapes of small asteroids and those of collisional fragments /6,16,17/. Actually, the mean asteroid amplitude (some 0.3 mag) is somewhat larger than that expected for typical fragments, which have three perpendicular axes in the ratios 1:0.7:0.5, yielding a mean amplitude of 0.2 mag when one assumes a triaxial ellipsoidal shape: but this discrepancy can be easily explained as a result of observational selection effects (against small lightcurve amplitudes), of complex amplitude-enhancing effects due light–scattering and shadowing effects on asteroid surfaces (especially when they are seen at nonzero phase angles), and also of the fact that the assumption of nearly ellipsoidal shapes for irregular fragments is a poor one.

Recent CCD lightcurve observation campaigns /18/ have extended the validity of the above conclusion down to main–belt asteroids a few km across. The average amplitude is nearly equal to that observed for asteroids ten or twenty times larger, confirming that a single process — collisional fragmentation — is determining the shapes of all bodies smaller than about 50 km. This is in agreement with the results of collisional evolution models /19/, that suggest that all asteroids in this size range should be multi–generation fragments rather than surviving original bodies. Actually, laboratory fragments also appear to maintain nearly constant mean shapes over a wide range of sizes, impact energies, and target strengths /20/. While the physical mechanisms responsible for this property are still unclear, it appears that they are effective also for asteroidal catastrophic collisions.

GASPRA

951 Gaspra, the first asteroid ever approached by a space probe /21/, has a strongly irregular global shape, with a mean radius of about 7 km but triaxial dimensions of 19 by 12 by 11 km. The

shape cannot be matched well by a triaxial ellipsoid, due to the presence of very large concavities, bulges and ridges; however, the general appearance of the surface is fairly smooth, with no sharp edges and a relative scarcity of km–sized craters. Several linear depressions are apparent, possibly similar to Phobos' grooves and related to internal fractures. Subtle colour and albedo variations hint to the presence of surface regolith, which was unexpected for such a small body.

A possible explanation of these observations is that Gaspra is a rubble pile, overlain by a substantial regolith layer; the large concavities would represent giant impact scars, formed by events close to the catastrophic break–up threshold. Collisional models /22/ show that even for bodies as small as Gaspra a large fraction of the material ejected after impacts can be reaccumulated by self–gravity both in the shattering and in the cratering regime. Thus Gaspra may have been originated in a shattering event followed by the reaggregation of a few large blocks; the subsequent formation of a few giant craters may have produced the surface regolith and erased the small–scale topography.

The first image of Gaspra transmitted by the *Galileo* probe has also been used as a testing ground for checking three different lightcurve inversion methods, widely used in the past to derive asteroid shapes and polar direction /23/. The results have been very positive: while lightcurve data can provide information only on the global morphology, all the methods have given a fairly accurate fit when compared with the actual Gaspra image. This is remarkable, since Gaspra's shape is so irregular that the lightcurves show a number of complex features (i.e., asymmetric extrema), which have to be fitted by non–ellipsoidal models of the surface. Earth–based observations will remain for a long time the main source of physical data for improving our understanding of the general asteroid population, even while some individual objects will be studied *in situ* by space missions; therefore, it is essential to exploit the latter also for putting on a more solid ground the models and theories on asteroid properties derived by remote sensing techniques.

FAMILIES, OUTER–BELT AND NEAR–EARTH ASTEROIDS

Nowadays, the asteroids are no more sees as a uniform population of rocky chunks, with orbits scattered in a wide belt between Mars and Jupiter and more or less homogeneous physical properties. The asteroid complex is instead divided into a number of sub–populations, filling different dynamical niches and with strong correlations between orbital and physical characteristics. The shapes are not an exception: a number of recent observations and surveys, aimed at particular sub–populations, have found intriguing hints that the average asteroid shape may depend strongly on the dynamical and collisional environment, and therefore may give important insight on the origin, history and interrelations of these bodies. Here we shall just give some examples of these recent findings, without going into details — also because the explanations proposed so far are in most cases just speculative. Our list includes the following items: (1) the Eos and Koronis dynamical families display different distributions for lightcurve amplitudes, which may be due either to different average shapes or to non–isotropic distributions of the polar axes /24/; (2) outer belt and Trojan asteroids have an incidence of high lightcurve amplitudes larger than that of main–belt asteroids of similar size /25/; (3) several km–sized near–Earth asteroids, observed by the radar techniques, have revealed to be contact binaries /26/. In all the three cases, the physics of the collisional break–up processes occurring under different conditions are probably responsible for the observed anomalies — but this physics is still poorly understood, and a progress in this field will certainly have a decisive role in the future understanding of the small bodies of the solar system.

REFERENCES

1. B. Bertotti and P. Farinella, *Physics of the Earth and the Solar System*, Kluwer, Dordrecht, 1990.

2. P. Farinella, Small satellites, in: *The Evolution of the Small Bodies of the Solar System*, eds. M. Fulchignoni and Ľ. Kresák, North–Holland, Amsterdam 1987, p. 276.

3. S.K. Croft, Proteus: Geology, shape, and catastrophic destruction, *Icarus* 99, 402–419 (1992)

4. V. Zappalà, Physical and statistical interpretations of asteroid lightcurves, in: *The Evolution of the Small Bodies of the Solar System*, eds. M. Fulchignoni and Ľ. Kresák, North–Holland, Amsterdam 1987, p. 91.

5. P. Magnusson, M.A. Barucci, J.D. Drummond, K. Lumme, S.J. Ostro, J. Surdej, R.C. Taylor and V. Zappalà, Determination of pole orientations and shapes of asteroids, in: *Asteroids II*, eds. R.P. Binzel, T. Gehrels and M.S. Matthews, Univ. of Arizona, Tucson 1989, p. 66.

6. R.P. Binzel, P. Farinella, A. Zappalà and A. Cellino, Asteroid rotation rates: Distributions and statistics, in: *Asteroids II*, eds. R.P. Binzel, T. Gehrels and M.S. Matthews, Univ. of Arizona, Tucson 1989, p.416.

7. S. Chandrasekhar, *Ellipsoidal Figures of Equilibrium*, Yale University, New Haven/London 1969.

8. A. Cellino, M. Di Martino, J. Drummond, P. Farinella, P. Paolicchi and V. Zappalà, Vesta's shape, density and albedo features, *Astron. Astrophys.* 219, 320–321 (1989).

9. R.L. Millis, O.G. Franz, L.H. Wasserman, L.A. Lebofsky, E. Asphaug, W.B. Hubbard, D.M. Hunten, M. A'Hearn, R. Schnurr, A.R. Klemola, W. Osborn, F. Vilas, A.E. Potter, P.D. Maley and P.L. Manly, Constraints on the size, shape, and density of (4) Vesta, *Bull. Amer. Astron. Soc.* 21, # 4, 1247 (1990).

10. R.L. Millis and D.W. Dunham, Precise measurement of asteroid sizes and shapes from occultations, in: *Asteroids II*, eds. R.P. Binzel, T. Gehrels and M.S. Matthews, Univ. of Arizona, Tucson 1989, p.148.

11. D.W. Dunham and 45 coauthors, The size and shape of (2) Pallas from the 1983 occultation of 1 Vulpeculae, *Astron. J.* 99, 1636–1662 (1990).

12. P. Farinella, P. Paolicchi, E.F. Tedesco and V. Zappalà, Triaxial equilibrium ellipsoids among the asteroids?, *Icarus* 46, 114–123 (1981).

13. P. Farinella, P. Paolicchi and V. Zappalà, The asteroids as outcomes of catastrophic collisions, *Icarus* 52, 409–433 (1982).

14. A. Cellino, M. Di Martino, P. Farinella, P. Paolicchi and V. Zappalà, Do we observe light curves of binary asteroids?, *Astron. Astrophys.* 144, 355-362 (1985).

15. J.D. Drummond, S.J. Weidenschilling, C.R. Chapman and D.R. Davis, Photometric geodesy of main–belt asteroids, II. Analysis of lightcurves for poles, periods, and shapes, *Icarus* 76, 19–77 (1988).

16. F. Capaccioni, P. Cerroni, M. Coradini, P. Farinella, E. Flamini, G. Martelli, P. Paolicchi, P.N. Smith and V. Zappalà, Shapes of asteroids compared with fragments from hypervelocity impact experiments, *Nature* 308, 832–834 (1984).

17. V. Catullo, V. Zappalà, P. Farinella and P. Paolicchi, Analysis of the shape distribution of asteroids, *Astron. Astrophys.* 138, 464–468 (1984).

18. R.P. Binzel, S. Xu, S.J. Bus and E. Bowell, Small main–belt asteroid lightcurve survey, *Icarus* 99, 225–337 (1992).

19. D.R. Davis, S.J. Weidenschilling, P. Farinella, P. Paolicchi and R.P. Binzel, Asteroid collisional history: Effects on sizes and spins, in: *Asteroids II*, eds. R.P. Binzel, T. Gehrels and M.S. Matthews, Univ. of Arizona, Tucson 1989, p. 805.

20. F. Capaccioni, P. Cerroni, M. Coradini, M. Di Martino, P. Farinella, E. Flamini, G. Martelli, P. Paolicchi, P.N. Smith, A. Woodward and V. Zappalà, Asteroidal catastrophic collisions simulated by hypervelocity impact experiments, *Icarus* 66, 487–514 (1986).

21. M.J.S. Belton, J. Veverka, P. Thomas, P. Helfenstein. D. Simonelli, C. Chapman, M.E. Davies, R. Greeley, R. Greenberg, J. Head, S. Murchie, K. Klaasen, T.V. Johnson, A. McEwen, D. Morrison, G. Neukum, F. Fanale, C. Anger, M. Carr and C. Pilcher, Galileo encounter with 951 Gaspra: First pictures of an asteroid, *Science* 257, 1647–1652 (1992).

22. J.–M. Petit and P. Farinella, Modelling the outcomes of high–velocity impacts between small solar system bodies, *Celest. Mech.*, in press (1993).

23. M.A. Barucci, A. Cellino, C. De Sanctis, M. Fulchignoni, K. Lumme, V. Zappalà and P. Magnusson, Ground–based Gaspra modelling: Comparison with the first Galileo image, *Astron. Astrophys.* 266, 385–394 (1992).

24. R.P. Binzel, Collisional evolution in the Eos and Koronis asteroid families: Observational and numerical results, *Icarus* 73, 303–313 (1988).

25. V. Zappalà, M. Di Martino, A. Cellino, P. Farinella, G. De Sanctis and W. Ferreri, Rotational properties of outer belt asteroids, *Icarus* 82, 354–368 (1989).

26. S.J. Ostro, J.F. Chandler, A.A. Hine, K.D. Rosema, L.L. Shapiro and D.K. Yeomans, Radar images of asteroid 1989 PB, *Science* 248, 1523–1528 (1990).

Pergamon

PII: S0273-1177(96)00011-7

Adv. Space Res. Vol. 19, No. 1, pp. (1)187–(1)194, 1997
© 1997 COSPAR
Printed in Great Britain. All rights reserved
0273–1177/97 $17.00 + 0.00

COMET HALLEY'S NUCLEUS: A PHYSICAL INTERPRETATION

H. U. Keller and N. Thomas

Max-Planck-Institut für Aeronomie, D-3411 Katlenburg-Lindau, Germany

ABSTRACT

The nucleus of comet Halley was the first to be detected and observed by cameras on board spacecraft during the fly-bys in 1986. The interpretation of both these observations and of the results from other experiments as well as many subsequent corroborative measurements of comet Halley and of other comets are reviewed. Cometary nuclei are relatively large and non-spherical in shape. They show activity only on small fractions of their surfaces. The physical processes leading to such heterogeneous surfaces are not yet understood. Our present concept of cometary nuclei and their formation are discussed. Open questions and problems are addressed.
©1997 COSPAR. All rights reserved

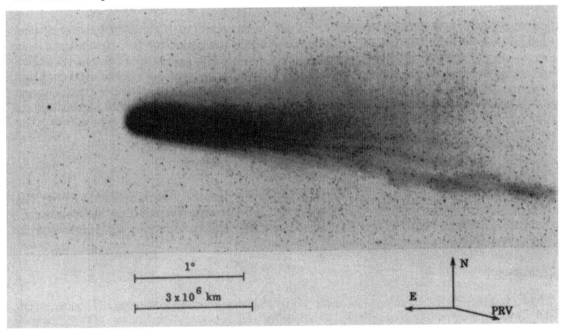

Fig. 1. Comet Halley observed from Chile shortly after the Giotto encounter. The ion tail shows more structure than the dust tail. (Image taken by F. Miller, University of Michigan)

INTRODUCTION

Ten years ago comets would not have been included in a meeting on *The Shapes and Gravitational Fields of the Planets and Smaller Bodies of the Solar System*. The nature and even the existence of their nuclei was still hypothetical. Only the fly-bys of comet Halley in 1986 by the VEGA and Giotto spacecraft disclosed the nucleus of this most famous comet. The pictures taken by the on board cameras /1,2/ confirmed the predictions of a solid body in the centre of the cometary coma postulated by Whipple /3/ more than 30 years earlier. Small bodies of dimensions in the order of ten kilometres are the sources of spectacular phenomena such as the gas coma, the ion and dust tails. Their manifestations reach dimensions many orders of magnitude larger than the nucleus and can exceed the volume of any other phenomenon in the Solar System (Fig. 1). The visible coma (sometimes called the cometary head) typically extends out to 10^5 km. Molecules subliming from the nucleus are dissociated, ionised and finally swept away by the supersonic solar wind. The ion tail of an active comet extends for several 10^7 km in the antisolar direction. Dust grains are dragged from the nucleus by the gas. The fine particles scatter the solar radiation and usually provide the strongest contribution to the coma emission. The dust grains are driven away from the nucleus by radiation pressure. After perihelion the curvature of the dust tail often reflects the fast change of true anomaly of the cometary orbit.

COMETS - WITNESSES OF SOLAR SYSTEM FORMATION

What makes comets so interesting that several space missions were dedicated to close-up investigations of Halley's comet? It is their unique activity that demonstrates that volatile compounds comprise a large fraction of their volume. These volatiles could not survive in the inner Solar System for an extended time. The observed short ($P < 200$ y) and long period (10^6 y and longer) period comets have to be replenished from a reservoir far from the Sun. The Oort cloud /4/ of more than 10^{12} comets extends from about 10^4 AU out to a distance comparable to that of neighbouring stars and was formed while the planetary system was accreted. New comets are being scattered into the inner Solar System by passing stars, large molecular clouds, and galactic tides. They are the most pristine objects in our Solar System. Created at low temperatures at the fringes of the forming planetary system, too small to be altered by internal heat produced by their own gravitational accretion, and stored far away from the Sun at very low temperatures. Most of the short period comets may come from a second reservoir extending from the outer fringes of the planetary system to about 10^4 AU, the Kuiper belt. Model calculations show that the Oort cloud itself may not be stable over the lifetime of the Solar System (4.5 10^9 y) and may have been replenished by comets originating in the Kuiper belt (see /5/ for a review).

Comets supposedly carry information about the physical conditions of the outer Solar System at the time when the planets formed. They should still reflect the chemical composition of the solar nebula, possibly even carrying pristine interstellar compounds and dust particles. This unique role of comets make them a prime target for investigations related to questions of the formation of planetary systems and composition of interstellar and planetary molecular clouds. While studying comets and their interaction with solar photons and particle radiation is also highly interesting and in many ways unique the ultimate goal of cometary research goes beyond the physical nature of comets themselves to the questions of their origin and the circumstances of their formation.

HIGHLIGHTS FROM THE COMET HALLEY FLY-BYS

Many physical properties of the nucleus were disclosed by the cameras during the spacecraft fly-bys. The imaging data were supplemented by observations of other space-borne and ground-based instrumentations. The results have refined our knowledge of cometary physics but as one would expect also changed our perception including aspects related to comet formation.

Composition

In situ mass spectrometry confirmed that water (H_2O) is the most abundant compound of comets, in fact, the dominating species constituting 80 % of the volatile material found in the coma. CO follows with 15 %. It is mostly (ca. 2/3) released from an extended source, the dust grains, rather than directly from the nucleus (about 1/3 or 5 %) /6/. All other compounds contribute only in the range of a few percent or less. 1/3 of the non-volatile component is made from organic material. Some dust grains are composed only from C, H, O, and N (CHON particles /7/), i. e. Polyoxymethylene (POM) was tentatively identified /8/. The particle size distribution covers many orders of magnitude ranging from the size of large macro molecules to centimetres, most of the mass is ejected from the nucleus in form of large particles /9/. The dust to gas ratio has consistently been found to be larger than one /10/.

Nucleus

The least surprising but nevertheless the most important result is that the cometary nucleus exists as a single solid body. Comet Halley's nucleus is larger and therefore darker (albedo about 4%) than predicted, of irregular shape and its activity is concentrated on about 20% of the insolated surface. The surface temperature reaches values as high as 400 K /11/ and exceeds the water sublimation temperature (ca. 200 K) by far. This confirms that most of the surface is covered by non-volatile material. Water ice can only be found in the areas of activity.

Bulk Properties of the Nucleus

The shape of the whole nucleus (Fig. 2) is revealed on images of the Halley Multicolour Camera (HMC, on board the Giotto space-craft) because the unilluminated hemisphere is outlined against the stray light from dust particles behind the nucleus. Only this cir-cumstance makes it possible to determine the 3-dimensional shape and size of the nucleus in combination with images from the Vega fly-bys taken at different aspect angles (Table 1). The nucleus can be approximated by an almost symmetric ellipsoid of 15.3 km x 7.2 km x 7.22 km (± 0.5 km in all dimen-sions). The volume of a 3-dimensional model /12/ is 365 km^3 and its surface area is 294 km^2 (Fig. 3). Rickman /13/ has derived the mass of the nucleus from a study of the non-gravitational forces to 1 - 3·10^{14} kg; its density is found to be 550 (± 250) kg m^{-3}.

The geometric albedo of 0.04 (+0.02 -0.01) derived from Vega images /1/ assuming a Moon-like phase function is darker than that of the Martian moons Phobos or Deimos and makes comets one of the darkest known objects of the Solar System. The reflectivity at a phase angle of 107° (the approach angle of the Giotto spacecraft) is only 0.006 /2/. The colour of the surface is slightly reddish

Fig. 2. A composite image of the nucleus of comet Halley constructed from images taken by the Halley Multicolour Camera during the Giotto encounter. The resolution changes from about 300 m at "big end" of the nucleus to about 100 m at the area of strongest activity (the brightest point).

with a gradient of 6 (±3)% (100 nm)$^{-1}$ over the wavelength range of 440 to 810 nm, somewhat similar to a P-type asteroid. The dust particles in the main jets are slightly more reddish /14/. The dark and reddish surface and dust support the existence of organic material.

Three areas of major activity were observed or inferred, altogether about 5 areas could be localized. All the active areas comprise only about 10% of the total surface or about 20% of the surface illuminated during the Giotto fly-by.

The rotatio of the nucleus was first derived from Vega images to be 54 h around an axis perpendicular to the long axis of the nucleus. Ground-based observations revealed a variation of activity (in gas and less clearly in dust) of 7.4 d /16/. A more complex rotational motion that includes a slow motion around the long body axis and nutation (or free precession) has been first suggested by /17/. Recently /15/ tried to reconcile the spacecraft and ground-based observations deriving the motion defined in Table 1. The number of free parameters of such an excited state of rotation is large enough to achieve a reasonable self-consistent fit to the few limited space-borne observations and to the brightness variations of the coma. The orientation of the major dust jet of the model conflicts with interpretations of the HMC observations. The model does not explain the change of rotation period derived from modelling the breathing of the coma in March and April 1986 /18/.

The centrifugal forces at the tips of the nucleus almost compensate the gravitational attraction for rotation periods of a day or smaller due to the elongated shape (2:1 ratio) and the postulated very low density of the nucleus. The resulting low escape velocities and higher moments of the gravitational field facilitate the release of large debris from the nucleus.

TABLE 1 *Properties of the Nucleus of Comet Halley*

Comet Halley's Nucleus		
Projected shape (full outline)	max. length 14.2 ±0.3 km max. width 7.4 ±0.2 km	HMC
Model body	15.3 x 7.2 x 7.22 km³	/12/
Volume	420 ±80 km³ tri-axial ellipsoid 365 km³ model body	/12/
Surface	294 km²	
Topography	mountains, ridges, terraces	HMC
Activity	concentrated in 3 major areas, ≤ 10 % of surface	HMC
Geometric albedo	0.04 +0.02 -0.01	/1/
Colour (reddish)	reflectivity gradient: 6 ±3 % (100 nm)⁻¹ from 440 to 810 nm	/14/
Mass	1-3·10¹⁴ kg	from non-gravitational forces /13/
Density	550 ±250 kg m⁻³	/13/
Rotation (complex)	spin period 2.84 d 7.1 d around long axis 3.7 d nutation	/15/

RESULTS OF THE HALLEY MULTICOLOUR CAMERA

<u>Topography and Morphology of the Nucleus</u>

The fully visible outline of the nucleus shows a body the shape of which strongly deviates from a sphere and even from the best fit ellipse. Astonishingly the backside limb follows a straight line over nearly 10 km in length to be terminated by an almost rectangular corner (as seen by HMC in projection), the *Duck Tail*, protruding out to ΔR/R = 0.3 where R is the radius of the best fit ellipse at the point of the feature. The relatively smooth course of the terminator north and south of the *Central Depression* being paralleled by a bright band (*Ridge*) at the morning (southern) side indicates large scale features such as

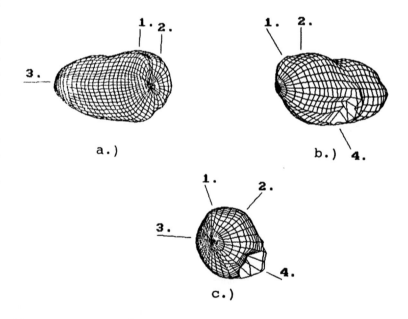

Fig. 3. A wire model of the nucleus of comet Halley derived from spacecraft observations /12/. The viewing angles are: c) the longest axis points towards the reader, a) and b) it is rotated 40° to the right and left, respectively.

a terrace. The *Central Depression* (see Fig. 4 for locations of the features) tapers towards the *Mountain*. Its illuminated tip is estimated to extend to $\Delta R/R = 0.25$ or 900 m above the best fit ellipsoid.

These large scale structures are counterbalanced by structures with typical dimensions of the order 500 m to 1 km. The most obvious example is seen at the *Chain of Hills*. This scale length dominates even at places where the resolution elements of the camera images are smaller (down to ca. 100 m) near the active region and inside the *Crater*. This roundish feature is clearly visible even in the raw HMC data. It covers a projected area of 12 km^2 near the illuminated limb of the nucleus and is therefore be distorted by the viewing geometry. The observed topographic features could be shallow (the depth of the *Crater* was estimated to about 200 m /19/) since the solar illumination is very slanted.

Evolution of the Surface

The sublimation of the water ice leads to a loss of 5 to 20 m of dust/ice material above active areas (about 2 m average over the whole surface) during each orbit. For the last 30 recorded apparitions of comet Halley a depth comparable to that of the *Crater* is reached. The dynamical evolution of the meteor streams associated with comet Halley indicates an age of 2000 to 3000 revolutions on the comet's present orbit. The mass of the meteor streams seems several times larger than the mass of the present nucleus /20/.

The above estimates make it clear that we are looking at a very evolved body showing strong erosion. The presently observed surface was the interior of the original nucleus that was several times more massive. The low density of the nucleus and therefore its low surface gravity and its irregular shape (by virtue of the accretion process or by sublimation of heterogeneous nucleus material) lend themselves to the concept that protrusions are broken away and could escape. Even splitting (commonly observed) of cometary nuclei seems then a plausible consequence.

Comet Halley's nucleus neither looks like being made out of "snow" /21/ nor does it look like a shrinking "ball". Erosion by sublimation of a predominantly volatile material should smooth the surface and flatten the protrusions in contrast to several topographic features (*Mountain*, *Duck Tail*) observed on the surface of comet Halley.

At close range the nucleus looks rather inactive, only about 10 % of its surface shows activity which concentrates the dust into jets. Once this relative inactivity of comet Halley was established ground-based observations with modern instrumentation showed that typical short period comets display even less activity, in the range of 1 % of their surface /22/. Most of the surface of cometary nuclei is covered by non-volatile material, a mantle or a crust? The small overall variations of the surface reflectivity (less than 50 %) indicate that active areas (the interior of the nucleus (?)) are not qualitatively different from the surface.

The Structure of Cometary Nuclei

The typical scale-length of the topographic features (500 m to 1 km) indicates that the nucleus could have formed from subnuclei of a typical size of 1 km by non-destructive collisions rather than being directly accreted from small dust grains. The concept of a fractal-like growth (although this concept strictly works only for small particles) requires low relative velocities of the colliding subnuclei, not exceeding a few metres per second /23/. Even then the low density subnuclei would penetrate each other creating a physically inhomogeneous body with regions of higher densities and nearly void zones in between the subnuclei. These boundaries could be connected to the areas of activity and certainly represent weak elements along which cometary nuclei would preferably split.

As discussed above, the nuclear material does not look "snow-like" with the volatile (icy) component dominating the physical consistency. The detailed observations of comet Halley lead to a picture of the nucleus that is shifting away from the predominance of the volatile (water) component contaminated with dust to a nucleus that is made from a matrix of non-volatile material with a strong component of volatiles (mainly water ice) /24/.

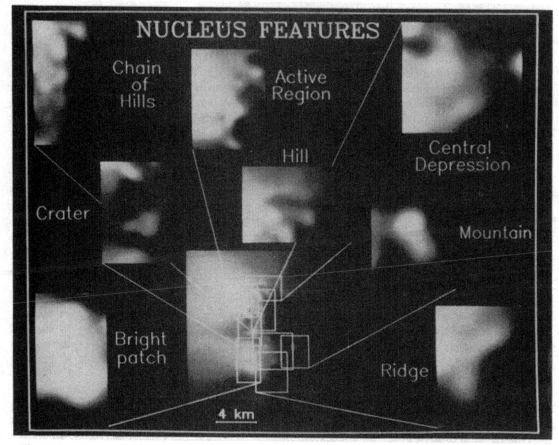

Fig. 4. Several topographic features of the nucleus observed by HMC. Their contrast is individually stretched in the enlarged cut-outs.

Models of the formation of dust mantles on a volatile nucleus show that already very thin mantles in the order of 100 times the average pore size choke the sublimation. These mantles, however, are blown away during the following perihelion passage of the comet /25/. Only large (decimetre size) pieces could stay behind. This behaviour is not consistent with dust particles being imbedded in (water) ice but with the inference that the refractory material forms the matrix (structure) that determines the physical properties of the nucleus. This matrix of refractory material must be strong enough to keep large particles intact. Fireball meteoroids (with a typical diameter in the order of 1 m and of very low density /26/) survive many orbits around the Sun and are certainly completely depleted of all volatile material. One has, therefore, to conclude that the refractory component and not the volatile water ice determines the porous structure and physical behaviour of a cometary nucleus.

It is important to stress that refractory material relates to all compounds less volatile than water including a large portion of organic materials that sublime at higher temperatures. About 1/3 of the particulate material observed during the spacecraft encounters with comet Halley is composed of organic compounds /27/. A considerable part of the volatile and abundant elements (H, C, O, and N) seems to be locked in complex organic molecules that are less volatile than water ice rather than to be stored as ices (CO, CO_2, N_2, CH_4 etc.) in the nucleus. These organic compounds may originate in the original interstellar material out of which the Solar System was formed. It is questionable whether there was enough time to form these complex molecules during the formation of the planetesimals or cometesimals that could have taken less than 10^6 y /28/.

SUMMARY AND CONCLUSIONS

Cometary nuclei are much bigger than indicated (and previously assumed) by their overall activity levels. They are of low density and therefore very porous. They display a very dark, slightly reddish surface and interior consistent with refractory material containing large quantities of organic (polymeric) compounds. Water ice constitutes the main volatile component. Cometary nuclei are not covered by thin mantles of small dust particles but possibly by rather large (decimetre size) pieces of their depleted matrix material or they just display depleted matrix material. The physical structure of the nuclear interior is determined by the prevailing refractory component. To express it in a single sentence: comets resemble more "dirt balls" than "snowballs".

The surprisingly low activity level of the most active short period comet (Halley) and the even lower activity levels of other comets could give the impression of decay. This is of course not in agreement with the hardly changing repetitive behaviour of most periodic comets. It is unclear how this restricted activity can be maintain over many revolutions around the Sun. The understanding of the nature of this restricted but persistent activity could be an important key to the nature of cometary nuclei and their chemistry and physics.

REFERENCES

1. R.Z.Sagdeev, J. Blamont, A.A. Galeev, V.I. Moroz, V.D. Shapiro, V.I. Shevchenko, and K. Szegö, Vega spacecraft encounters with comet Halley, *Nature* 321, 259-262 (1986).

2. H.U. Keller, C. Arpigny, C. Barbieri, R.M. Bonnet, S. Cazes, M. Coradini, C.B. Cosmovici, W.A. Delamere, W.F. Huebner, D.W. Hughes, C. Jamar, D. Malaise, H.J. Reitsema, H.J. Schmidt, W.K.H. Schmidt, P. Seige, F.L. Whipple, and K. Wilhelm, First Halley Multicolour Camera imaging results from Giotto, *Nature* 321, 320-326 (1986).

3. F.L. Whipple, A Comet Model I. The Acceleration of Comet Encke, *Astrophys. J.* 111, 375-394 (1950).

4. J.H. Oort, The structure of the cloud of comets surrounding the solar system and a hypothesis concerning its origin, *Bull. Astron. Inst. Neth.* IX, 91-110 (1950).

5. P.R. Weissman, Dynamical History of the Oort Cloud, in: *Comets in the Post-Halley Era*, ed. R. Newburn and J. Rahe, IR. Newburn and J. Rahe, Kluwer Academic Publishers, Dordrecht, The Netherlands, 1990, p. 463.

6. D. Krankowsky and P. Eberhardt, Evidence for the Composition of Ices in the Nucleus of Comet Halley, in: *Comet Halley, Investigations, Results, Interpretations Volume 1: Organization, Plasma, Gas*, Ed. J.W. Mason, Ellis Horwood Library, Chichester, 1988, p. 273.

7. J. Kissel, D.E. Brownlee, K. Büchler, B.C. Clark, H. Fechtig, E. Grün, K. Hornung, E.B. Igenbergs, E.K. Jessberger, F.R. Krueger, H. Kuczera, J.A.M. McDonnell, G.M. Morfill, J. Rahe, G.H. Schwehm, Z. Sekanina, N.G. Utterback, H.J. Völk, and H.A. Zook, Composition of Comet Halley dust particles from Giotto observations, *Nature* 321, 336-337 (1986).

8. W.F. Huebner, First Polymer in Space Identified in Comet Halley, *Science* 237, 628-630 (1987).

9. J.A.M. McDonnell, J. Kissel, E. Grün, R.J.L. Grard, Y. Langevin, R.E. Olearczyk, C.H. Perry, and J.C. Zarnecki, Giotto's Dust Impact Detection System DIDSY and Particulate Impact Analyser PIA: Interim Assessment of the Dust Distribution and Properties within the Coma, in: *20th ESLAB Symposium on the Exploration of Halley's Comet*, Eds. B. Battrick, E.J. Rolfe, and R. Reinhard, ESA-SP 250 part II, ESA, Publications Division, ESTEC, Noordwijk 1986, p. 25.

10. J.A.M. McDonnell, S.F. Green, E. Grün, J. Kissel, S. Nappo, G. Pankiewicz, and C.H. Perry, In-situ exploration of the dusty coma of comet P/Halley at Giotto's encounter: flux rates and time profiles from 10^{-19} kg to 10^{-5} kg., *Adv. Space Res.* 9, (3)277-(3)280 (1989).

11. C. Emerich, J.M. Lamarre, V.I. Moroz, M. Combes, N.F. Sanko, Y.V. Nikolsky, F. Rocard, R. Gispert, N. Coron, J.P. Bibring, T. Encrenaz, and J. Crovisier, Temperature and size of the nucleus of comet P/Halley deduced from IKS infrared Vega-1 measurements, *Astron. Astrophys.* 187, 839-842 (1987).

12. E. Merényi, L. Földy, K. Szegö, I. Tóth, and A. Kondor, The Landscape of Comet Halley, *Icarus* 86, 9-20 (1990).

13. H. Rickman, L. Kamél, M.C. Festou, and Cl. Froeschlé, Estimates of masses, volumes and densities of short-period comet nuclei, in: *Symposium on the Diversity and Similarity of Comets*, Eds. E.J. Rolfe and B. Battrick, ESA-SP 278E.J. Rolfe and B. Battrick, ESA Publications Division, ESTEC, Noordwijk, 1987, p. 471.

14. N. Thomas and H.U. Keller, The colour of comet P/Halley's nucleus and dust, *Astron. Astrophys.* 213, 487-494 (1989).

15. M.J.S. Belton, W.H. Julian, A.J. Anderson, and B.E.A. Mueller, The Spin State and Homogeneity of Comet Halley's Nucleus, *Icarus* 93, 183-193 (1991).

16. R.L. Millis and D.G. Schleicher, Rotational period of comet Halley, *Nature* 324, 646-649 (1986).

17. Z. Sekanina, Nucleus of comet Halley as a torque-free rigid rotator, *Nature* 325, 326-328 (1987).

18. M.R. Combi and U. Fink, P/Halley: Effects of Time-Dependent Production Rates on Spatial Emission Profiles, *preprint*(1992)

19. G. Schwarz, H. Craubner, W.A. Delamere, M. Goebel, M. Gonano, W.F. Huebner, H.U. Keller, J.R. Kramm, E. Mikusch, H.J. Reitsema, F.L. Whipple, and K. Wilhelm, Detailed analysis of a surface feature on comet P/Halley, *Astron. Astrophys.* 187, 847-851 (1987).

20. D.W. Hughes, P/Halley dust characteristics: a comparison between Orionid and Eta Aquarid meteor observations and those from the flyby spacecraft, *Astron. Astrophys.* 187, 879-888 (1987).

21. F.L. Whipple, The Cometary Nucleus - Current Concepts, in: *20th ESLAB Symposium on the Exploration of Halley's Comet*, Eds. B. Battrick, E.J. Rolfe, and R. Reinhard, ESA-SP 250 part II, ESA, Publications Division, ESTEC, Noordwijk 1986, p. 281.

22. M.F. A'Hearn, Observations of Cometary Nuclei, *Ann. Rev. Earth Planet. Sci.* 16, 273-293 (1988).

23. B. Donn, The origin and structure of icy cometary nuclei, *Icarus* 2, 396-402 (1963).

24. H.U. Keller, The Nucleus, in: *Physics and Chemistry of Comets*, ed. W.F. Huebner, Springer-Verlag, Berlin Heidelberg New York London Paris Tokyo Hong Kong Barcelona, 1990, p. 13.

25. F.P. Fanale and J.R. Salvail, An idealized short period comet model: surface insolation, H_2O flux, dust flux and mantle development, *Icarus* 60, 476-511 (1984).

26. Z. Ceplecha and R.E. McCrosky, Fireball end heights: A diagnostic for the structure of meteoric material, *J. Geophys. Res.* 81, 6257-6275 (1976).

27. J. Kissel, R.Z Sagdeev, J.L. Bertaux, V.N. Angarov, J. Audouze, J.E. Blamont, K. Büchler, E.N. Evlanov, H. Fechtig, M.N. Fomenkova, von H. Hoerner, N.A. Inogamov, V.N. Khromov, W. Knabe, F.R. Krueger, Y. Langevin, V.B. Leonas, A.C. Levasseur-Regourd, G.G. Managadze, S.N. Podkolzin, V.D. Shapiro, S.R. Tabaldyev, and B.V. Zubkov, Composition of comet Halley dust particles from Vega observations, *Nature* 321, 280-282 (1986).

28. S.J. Weidenschilling, Formation processes and timescales for meteorite parent bodies, in: *Meteorites and the early solar system*, eds. J.F. Kerridge and M.S. Matthews, University of Arizona Press, Tucson, 1988, p. 348.

AUTHOR INDEX